T0228695

Radar Principles
for the Non-Specialist

The SciTech Radar and Defense Series

How to Speak Radar by Arnold Acker
2001 / CDROM / ISBN 1-891121-17-0

Radar Principles for the Non-Specialist, 3rd Ed., by J.C. Toomay and P. J. Hannen
2004 / Softcover / 270 pages / ISBN 1-891121-28-6

Introduction to Airborne Radar, 2nd Ed., by George Stimson
1998 / Hardcover / 592 pages / ISBN 1-891121-01-4 or CDROM 1-891121-14-6

Understanding Radar Systems, by Simon Kingsley and Shaun Quegan
1999 / Hardcover / 392 pages / ISBN 1-891121-05-7

Air and Spaceborne Radar Systems by P. Lacomme et al.
2001 / Hardcover / 496 pages / ISBN 1-891121-13-2

Radar Design Principles, 2nd Ed., by Fred E. Nathanson
1998 / Hardcover / 656 pages / ISBN 1-891121-09-X

Radar Foundations for Imaging and Advanced Concepts by Roger Sullivan
2004 / Softcover / 475 pages / ISBN 1-891121-22-7

Understanding Synthetic Aperture Radar Images, by Chris Oliver and Shaun Quegan
2004 / Hardcover / 512 pages / ISBN 1-891121-31-6

Introduction to RF Stealth by David Lynch, Jr.
2004 / Hardcover / 570 pages / ISBN 1-891121-21-9

Radar Cross Section, 2nd Ed. by E. Knott, J. Shaeffer, and M. Tuley
2004 / Softcover / 611 pages / ISBN 1-891121-25-1

Introduction to Adaptive Arrays by Robert Monzingo and Thomas Miller
2004 / Softcover / 543 pages / ISBN 1-891121-24-3

Low Angle Radar Land Clutter by J. Barrie Billingsley
2002 / Hardcover / 700 pages / ISBN 1-891121-16-2

Airborne Early Warning System Concepts by Maurice Long
2004 / Softcover / 538 pages / ISBN 1-891121-32-4

Microwave Passive Direction Finding by Stephen Lipsky
2004 / Softcover / 320 pages / ISBN 1-891121-23-5

Digital Techniques for Wideband Receivers, 2nd Ed. by James Tsui
2004 / Softcover / 608 pages / ISBN 1-891121-26-X

Visit the SciTech website for details and current pricing
www.scitechpub.com

Radar Principles for the Non-Specialist

J. C. Toomay

Paul J. Hannen
Senior Systems Engineer
Science Applications International Corporation
Adjunct Professor, Department of Electrical Engineering
Wright State University
Fairborn, Ohio

Third Edition

SciTECH
PUBLISHING, INC.

Published by SciTech Publishing Inc.
Raleigh, NC
www.scitechpub.com

President and CEO: Dudley R. Kay
Copy Editing: Cavanaugh Editorial Services
Page Composition: J. K. Eckert & Company, Inc.
Cover Design: Brent Beckley

Distributed to Resellers in the U.S. and Canada by:

William Andrew Publishing
13 Eaton Ave.
Norwich, NY 13815
(800) 932-7045
www.williamandrew.com

Printed in the U.S.A.

10   9   8   7   6   5   4   3   2

This book is available at special quantity discounts to use as premiums and sales promotions, or for use in corporate training programs. For more information and quotes, please contact:

Director of Special Sales
SciTech Publishing, Inc.
7474 Creedmoor Rd. #192
Raleigh, NC 27613
Phone: (919) 866-1501
E-mail: sales@scitechpub.com
http://www.scitechpub.com

ISBN 1-891121-28-6 softcover

ISBN 1-891121-34-0 hardcover

To Kim, Amanda, and Adam

Special thanks to my colleagues, Mr. Keith Carter, Mr. Bruce Esken, and Mr. Mike Sutton, for their countless suggestions and review of the manuscript. Also, thanks to my family, relatives, friends, colleagues, students, and anyone who at one time or another said, "You should just write your own book." I am not there yet, but a lot closer.

Contents

Preface

WHAT THIS BOOK IS

This book is about radar. It will teach you the essentials of radar—the underlying principles. It is not like an engineering handbook that provides detailed design equations without explaining either derivation or rationale. It is not like a graduate school textbook that may be abstruse and esoteric to the point of incomprehensibility. Moreover, it is not like an anthology of popular magazine articles that may be gaudy but superficial. It is an attempt to distill the very complex, rich technology of radar into its fundamentals, tying them to the laws of nature on one end and to the most modern and complex systems on the other.

WHO IT'S FOR

If you are intellectually curious about "the way radars work," not satisfied with a casual explanation, yet without the time to take a course leading to a Master's degree, this book will provide you the level of comprehension you seek. If you have taken a radar short course but find you need a better foundation and/or need the concepts tied together better, this book will do that.

If your work requires you to supervise or meet as coequals with radar systems engineers or designers, this book will allow you to understand them, to question them intelligently, and perhaps to provide them with a perspective (a dispassionate yet competent view) that they lack.

If you are trained in another discipline but have been made the manager of a radar project or a system program that has one or more radars as subsystems, this book will provide you with the tools you need, not only to give your team members confidence but also to make a substantive technical contribution yourself.

For over 20 years, I was associated with scientists, engineers, mathematicians, business administration graduates, and other college-educated people, all of whom aspired to jobs with broader responsibilities. Without exception, these people craved expertise in the various technical areas their expanding horizons exposed them to. One of the principal technical areas was radar. (Surprisingly, even the electrical engineers with recent degrees had no knowledge of radar, which is really a specialty to be taken up in graduate school.) A rudimentary version of this book was put together in response to that particular situation, but my subsequent experience revealed an even broader need for knowledge of this sort.

HOW IT APPROACHES RADAR

Because this book is focused on imparting whole pieces of knowledge, developed with an evolutionary approach and tied together with a thread of logic, it starts with electromagnetic propagation, describes a radar of the utmost simplicity, and derives the radar range equation from that simple radar. Once the range equation is available, the book attacks the meaning of each term in it, moving through antennas, detection and tracking, radar cross section, waveforms and signal processing, and systems applications in an orderly progression. At the finish, the reader should be able to complete an acceptable, first-order radar design, including some trading-off of design parameters to achieve a more efficient system. However, more importantly, the reader should know enough to critique the designs of others and to understand which technical issues are fundamental and which are simply design frills. While clever design ideas, and acronyms for them, are rampant, radar functions do not change. Thus, while there is no way to keep up with each new design wrinkle, knowledge of the principles that govern them will give instant critical understanding.

WHAT IS UNUSUAL

This book does three things that—although perhaps not unique—are unusual. It presents a comprehensive set of radar principles, including all the features of the latest applications, in a relatively short volume. It presents these principles with their underlying derivations, using the simplest mathematics possible, explaining the steps and using only popularly tabulated functions, integrals, and other expressions. In addition, it uses the same method of derivation, the same mathematics, and the same conceptual approach to discussing all antennas and waveforms. Other authors are aware of the analogs between spectral functions and time functions, and often cite them, but they do not usually try to develop these concepts as a single whole.

WHAT IS USEFUL

This book presents information in logical chunks that are meant to be self-contained. Most chapters stand alone; the reader may be selective and still benefit. The chapters are scaled to their information content rather than the time required to absorb them. Some readers will require much more effort than others to master a particular chapter. Two levels of comprehension are provided: The reader may simply memorize key relationships, which are always identified in the text, or may master the principle and its derivation. Useful references are provided, with general references being preferred. Exercises at the end of each chapter are calculated not to stump the reader but to reinforce the concepts presented and illustrate their applications. Nevertheless, with the exception of chapters on waveforms and systems applications, most readers will find an evening's time sufficient for a chapter.

HOW IT IS ORGANIZED

The book goes from the fundamental toward the more complex and from philosophy to quantification. Its foundation is the radar range equation, as the following outline shows:

Chapter	Title	Process
1	Elementary Electromagnetics and the Radar Range Equation	History, technical fundamentals, deriving the radar range equation
2	Antennas	Treating terms in the radar equation
3	Detection and Tracking	Treating terms in the radar equation
4	Radar Cross Section	Treating terms in the radar equation
5	Waveforms and Signal Processing	Treating terms in the radar equation
6	Electronic Countermeasures (ECM)	How radar systems are countered
7	System Applications	Using the radar equation
8	Loose Ends of Radar Lore	Cleaning up loose ends
9	Radar Potentials and Limitations	Generalizing on what we have learned

J. C. Toomay

What Is New in the Third Edition?

The third edition builds on the unique style and solid foundation of the previous editions. Every section of each chapter was revised in some way. Many revisions are very noticeable, whereas some are subtle. Some of the more noticeable enhancements are the additions of equation numbers; more numerical examples, tables, and figures showing many of the concepts numerically; and exercises for almost all the concepts. These enhancements are aimed at making the book easier to learn from (and teach out of, which I do). Clarifications, enhancements, and additions are most prevalent in the Antenna, Detection and Tracking, Waveforms and Signal Processing, ECM, and System Applications chapters. The answers to the exercises at the end of each chapter have been double-checked and Appendix 4 updated. A solution set (Mathcad® and RTF files) for the vast majority of the exercises is available from the publisher.

Paul J. Hannen

Symbols

a	= Radius of the sphere, meters; Dimension of the square, meters
A	= Physical area of the antenna, meters; Phasor amplitude, volts; Effective aperture area, square meters; Effective intercept area of the target, square meters
$A(y)$	= Current distribution in the y-z plane
A_c	= Area of the clutter intercepted by the radar pulse, square meters
A_e	= Effective area of the receive antenna, meters; Effective area of the antenna, meters; Electromagnetic area of the object as seen by the radar, square meters
a_{nt}	= Radius of the nose tip, meters
$\mathbf{B}$	= Entire phase progression across the array, radians
B	= Receiver filter bandwidth, hertz; Number of different phase shifts, no units
B_j	= Jammer noise bandwidth, hertz
B_{RWR}	= RWR bandwidth, hertz
c	= Speed of light, 3×10^8 meters/second
$C(f)$	= Clutter voltage—frequency domain
C_0	= Clutter amplitude
C_{CR}	= Clutter cancellation ratio
d	= Array element spacing, meters; Physical distance between array elements, meters; Diameter of the cylinder, meters; Physical length of the antenna, meters
D	= Antenna length, meters; Length of the array, meters; Dimension of the object, meters; Synthetic aperture length, meters
dB	= Decibel, no units
dBm	= Decibels relative to one milliwatt, carries the units of milliwatts
dBsm	= Decibels relative to one square meter, carries the units of square meters
dBW	= Decibels relative to one watt, carries the units of watts
D_{eff}	= Maximum effective synthetic aperture length, meters
d_i	= Distance of the ith array element from the start of the array, meters
d_t	= Duty cycle, no units

E	=	Energy in one pulse, watt-seconds or joules; RMS signal energy in one pulse, watt-seconds or joules
$E(\theta)$	=	E-field produced by the current distribution, perpendicular to $A(y)$, voltage antenna gain pattern, no units
E/N_0	=	Single pulse energy signal-to-noise ratio, no units
ERP_j	=	Jammer effective radiated power, watts
f	=	Transmitted frequency, hertz; Frequency of the laser light, hertz
$f(t)$	=	Time-domain representation of the voltage pulse envelope, volts
$f(x)$	=	Probability of x occurring for the exponential probability distribution; Probability distribution
f_0	=	Frequency corresponding to the beam pointing at broadside, hertz
f_c	=	Radar carrier frequency, hertz
f_d	=	Doppler shift, hertz
f_{du}	=	Unambiguous Doppler shift—positive and negative Doppler shifts, hertz
f_{du+}	=	Unambiguous Doppler shift—positive Doppler shifts only, hertz
F_n	=	Receiver noise figure, no units
F_{RWR}	=	RWR noise figure, no units
g	=	Antenna gain of each element, no units
G	=	Radar antenna gain, no units; Boresight antenna gain for an array antennas, no units; Reradiation gain in the direction of the radar, no units; Radar receive mainbeam antenna gain, no units; Gain back in the direction of the lidar, no units
$G(f)$	=	MTI voltage frequency response
$G(\theta)$	=	Antenna gain pattern (power) for a uniform current distribution, no units
$g(\omega)$	=	Frequency spectrum of the pulse, volts
$G(\omega)$	=	Frequency spectrum of the pulse, watts
G_I	=	Integration gain, no units
G_j	=	Jammer transmit antenna gain, no units
G_p	=	Signal processing gain, no units
$G_p(f)$	=	Single-delay MTI power frequency response, no units
G_r	=	Receive antenna gain in the direction of the target, no units
G_{RWR}	=	RWR receive antenna gain, no units
G_{sl}	=	Radar receive antenna sidelobe gain in the direction of the jammer, no units

G_t = Transmit antenna gain in the direction of the target, no units

G_T = Transmit antenna gain, no units

g_x = Antenna gain of a half-wave dipole on a ground plane, no units

h = Planck's constant, 6.6×10^{-34} joule-seconds; Radar or target height, meters

h_r = Radar height, meters

h_t = Target height, meters

I = In-phase phasor of the signal, volts; Interference noise peak power, watts

J = Received jammer noise peak power, watts

$(J/S)_n$ = Multiple pulse jamming-to-signal ratio, no units

k = Boltzmann's constant, 1.38×10^{-23} watt-seconds/kelvins

k_r = Refraction factor, no units

l = Distance over which measurements are made, meters

L = Electrical distance between array elements, meters; Distance from the last measurement to the prediction point, meters; Length of the cylinder, meters

L_j = Jammer-related losses, no units

L_{RWR} = RWR system losses, no units

L_s = Radar system losses, no units; Bistatic radar system losses, no units

L_{tj} = Jammer transmit loss, no units

M = Number of times the detection threshold is crossed

n = Integer, $n = 0, 1, 2, \ldots$; Number of individual detections (scans or measurements) combined; Number of noise-like factors, no units; Integer number of the PRF, no units

$\bar{n}$ = Variance of fluctuations in the Poisson distribution, no units

$\bar{n}_s$ = Average number of signal photons arriving at the output of an optical receiver, no units

N = Receiver thermal noise, watts; Number of array elements, no units; Number of attempts; Number of chaff dipoles in the cloud, no units; Radar receiver and auxiliary receiver noise, watts

N_0 = RMS noise energy, watt/Hz or watt-seconds or joules

N_1 = One sample of receiver thermal noise power, watts

N_{bins} = Displacement in the number of range bins due to target Doppler, no units

N_e	=	Electron density, electrons/cm^2
N_f	=	Number of Doppler filters, no units
N_n	=	Receiver thermal noise power after integration of n_p noise samples, watts
n_p	=	Number of pulses integrated
$N\phi$	=	Number of phase coded segments, no units
O_n	=	Value of the nth observation
$\overline{O}_0$	=	New observation
p	=	Probability of crossing the detection threshold for each attempt
P	=	Peak transmit power, watts
$P(M,N,p)$	=	Probability of M detection threshold crossing out of N attempts
$P(mW)$	=	Power, milliwatts
$P(W)$	=	Power, watts
$P(x \geq T)$	=	Probability of x exceeding the value T
$P(x > T)$	=	Probability of noise exceeding a threshold, no units
P_{ave}	=	Average transmit power, watts
P_d	=	Probability of detection
P_{dc}	=	Cumulative probability of detection
P_{di}	=	Probability of detection for the ith individual detection
P_{fa}	=	Probability of false alarm
$P_{fa}(1)$	=	Probability of noise alone exceeding the detection threshold once (P_{fa})
$P_{fa}(2)$	=	Probability of noise alone exceeding the detection threshold twice
$P_{fa}(n)$	=	Probability of noise alone exceeding the detection threshold n times
P_i	=	Power in, watts
P_j	=	Jammer peak transmit power, watts
P_o	=	Power out, watts
P_r	=	Received energy, watts-seconds or joules
PRF	=	Pulse repetition frequency, hertz
PRI	=	Pulse repetition interval, seconds
Q	=	Quadrature phasor of the signal, volts
r	=	Radius of the flat back, meters
R	=	Range from the radar to the object, meters; Radar-to-target (and clutter) range, meters; Radar-to-target/jammer range, meters; Range to the outer edge of the SAR image, meters; Lidar-to-target range, meters

r_0	=	Radius of the sphere, meters
R_1	=	Measured radar-to-target range at time t_1, meters; Transmitter-to-target range, meters
$R_1 R_2$	=	Bistatic detection range product, square meters
R_2	=	Measured radar-to-target range at time t_2, meters; Receiver-to-target range, meters
R_3	=	Predicted radar-to-target range at time Δt in the future, meters
R_{bt}	=	Burnthrough range, meters
R_d	=	Radar detection range, meters
R_{dot}	=	Object-to-radar range rate, meters/second
R_{dotu}	=	Unambiguous range rate—positive and negative Doppler shifts, meters/second
R_{dotu+}	=	Unambiguous range rate—positive Doppler shifts only, meters/second
R_E	=	Radius of the Earth, 6371 kilometers
R_{ff}	=	Far field distance, meters
R_h	=	Range to the horizon, meters
R_{hc}	=	Clutter horizon, meters
R_{ht}	=	Target horizon, meters
R_j	=	Radar-to-jammer range, meters
R_{LOS}	=	Radar line of sight, meters
R_u	=	Unambiguous range, meters
r_x	=	Radius of curvature in the x-plane, meters
r_y	=	Radius of curvature in the y-plane, meters
S	=	Received signal power, watts; Target signal present in the radar receiver, watts
S/C	=	Signal-to-clutter ratio, no units
S/C_{MTI}	=	Signal-to-clutter ratio, no units
S/I	=	Single pulse signal-to-interference ratio, no units
$(S/I)_n$	=	Multiple pulse signal-to-interference ratio, no units
$(S/J)_n$	=	Multiple pulse signal-to-jamming ratio, no units
S/N	=	Single pulse signal-to-noise ratio, no units
$(S/N)_1$	=	Single pulse signal-to-noise ratio, no units
$(S/N)_n$	=	Signal-to-noise ratio after integration of n_p pulses, no units
$(S/N)_p$	=	Processed signal-to-noise ratio, no units

$(S/N)_{pc}$	= Post-cancellation signal-to-noise ratio, no units
$(S/N)_{RWR}$	= RWR single pulse radar signal-to-noise ratio, no units
S_1	= Received target signal power from one pulse, watts
S_n	= Received target signal power after integration of n_p coherent pulses, watts
SNR_d	= Signal-to-noise ratio required for detection (detection threshold), no units
S_{sl}	= Target signal present in the auxiliary receiver, watts
t	= Duration of the measurement, seconds
T	= Threshold value, watts; Time ahead the predication is made, seconds; Time it takes for the object to rotate one time, seconds; Pulse burst time, seconds
T/N	= Threshold-to-noise ratio, no units
T_0	= Receiver standard temperature (usually room temperature 290 K), kelvins
T_e	= Effective noise temperature, kelvins
T_{fa}	= Average time between false alarms, seconds
T_I	= Coherent integration time, seconds; Integration time, seconds
T_{ill}	= Target illumination time, seconds
T_{RWR}	= RWR system noise temperature, kelvins
T_s	= Time between measurements, seconds; Receiver system noise temperature, kelvins
T_t	= Time it takes the target to transit from the left edge to the right edge of the beam, seconds
v	= Collision frequency of the electrons with particles
V	= Volume of the sphere, cubic meters (m^3); Pulse amplitude, volts
V_{ac}	= Aircraft speed, meters/second
$V_{clutter}$	= Clutter range rate relative to the aircraft, meters/second
V_i	= Voltage in, volts
v_n	= nth blind speed, meters/second
$\hat{v}_n$	= Smoothed target velocity, meters/second
V_o	= Voltage out, volts
W	= Weighting function
X_n	= Measured target position, meters
X_{pn}	= Predicted target position, meters

$\bar{X}$	=	Current mean
$\bar{X'}$	=	Estimate
$\hat{x}_n$	=	Smoothed target position, meters
$\bar{X}_n$	=	Average after n observations
$\bar{X}_{n-1}$	=	Average after $n-1$ observations
Δv	=	Error in the velocity estimate, meters/second
$\Delta \phi$	=	Difference in phase, radians or degrees
$\Delta \lambda$	=	Wavelength excursion, meters
$\Delta \theta$	=	RMS angular error, radians; Angle resolution, degrees or radians
$\Delta \theta_0$	=	Maximum beam steering error, radians or degrees
Ω	=	Faraday rotation, degrees
α	=	Position smoothing parameter, no units; Grazing angle of the radar wave relative to the ground, radians or degrees; Attenuation, dB/Nmi
β	=	Velocity smoothing parameter, no units; Frequency excursion of the FM pulse, hertz; Bandwidth of a phase modulated pulse, hertz
β_i	=	Phase shift at the ith element, radians
δf_d	=	RMS Doppler accuracy, hertz
δR	=	RMS range accuracy, meters
δR_{dot}	=	RMS range rate accuracy, meters/second
δ_θ	=	Angle measurement accuracy, radians or degrees
ε	=	Arbitrary phase difference from center to edge of the antenna, meters
ε_I	=	Noise-like factors, square meters
ε_R	=	Range error at the prediction point, meters; Error in range due to unknown Doppler shift, meters
ε_θ	=	Angular error at the prediction point, radians
ϕ	=	Random polarization angle, radians; Phasor angle from the in-phase axis, radians or degrees
η	=	Quantum efficiency (the percentage of incident photons converted into electrons), no units; Index of refraction
λ	=	Wavelength, meters
λ_0	=	Wavelength corresponding to the beam pointing at broadside, meters

θ	= Mean and standard deviation of the exponential probability distribution; Polarization angle, radians; Peak to first null in the RCS pattern, radians; Angle from the aircraft velocity vector to the clutter, radians or degrees
θ_0	= Desired mainbeam pointing angle relative to the array boresight, radians; Beam steering angle, radians or degrees
θ_1	= Beam steering limits, radians or degrees
θ^2	= Variance of the exponential probability distribution; Effective aperture beam area, square radians
θ_{3dB}	= Antenna half-power (−3 dB) beamwidth
$\theta_{3dB}(\theta_0)$	= Array antenna half-power (−3 dB) beamwidth as a function of beam scan angle, radians or degrees
θ_{dot}	= Antenna scan rate, degrees/second or radians/second
θ_g	= Angle at which a grating lobe will appear, radians or degrees
θ_{nn}	= Antenna null-to-null beamwidth for a Uniform current distribution, radians or degrees
ρ	= Antenna efficiency, $0 < \rho < 1$, no units; Reflectivity of the target to the lidar wavelength, no units
σ	= Target radar cross section, square meters (m^2); Radar cross section for non-isotropic object, square meters; Radar cross section at a polarization angle, square meters
$\sigma(m^2)$	= Radar cross section, square meters (m^2)
$\bar{\sigma}$	= Average radar cross section across all polarization angles, square meters
$\bar{\sigma}$	= Average chaff radar cross section, square meters
$\bar{\sigma}_{HH}$	= Average radar cross section (horizontal transmit—horizontal receive), square meters
$\bar{\sigma}_{HV}$	= Average radar cross section (horizontal transmit—vertical receive), square meters
σ^2/n	= Variance of all measurements that have gone before
σ_0	= Radar cross section of a nondepolarizing target, square meters; Clutter reflectivity, m^2/m^2
σ_0^2	= Variance of the most current measurement
σ_b	= Target bistatic radar cross section, square meters

σ_c	=	Radar cross section of a large curved convex shape, square meters; Clutter radar cross section, square meters; Standard deviation of the clutter Doppler
σ_{cly}	=	Radar cross section of a cylinder, square meters
σ_d	=	Radar cross section of a dihedral corner reflector, square meters
$\sigma_{flatback}$	=	Radar cross section of the flat back, square meters
σ_{MTI}	=	Clutter radar cross section residue in a radar cell having clutter but no target, square meters
$\Sigma\sigma_i$	=	Sum of the reflection coefficients from all contributing scatterers, no units
$\sigma_{nosetip}$	=	Radar cross section of nose tip, square meters
σ_p	=	Peak chaff radar cross section, square meters
$\sigma_{Rayleigh}$	=	Radar cross section of a sphere in the Rayleigh region, square meters
σ_s	=	Radar cross section of a sphere, square meters
σ_t	=	Radar cross section of a trihedral corner reflector, square meters
σ_w	=	Radar cross section of a long wire, square meters
τ	=	Pulse duration, or pulse width, or pulse length, seconds
ω	=	Radial velocity of the antenna, radians/second; Radar frequency, radians/second
ω_o	=	Angular frequency the pulse, radians/second
ψ	=	Reflection coefficient, m^2/m^2

1

Elementary Electromagnetics and the Radar Range Equation

HIGHLIGHTS

- Some fundamentals of radio waves from Faraday, Maxwell, and Hertz
- Putting together a simple radar and discussing its principal parts
- Deriving the radar range equation from first principles
- Discovering special features of surveillance and tracking radars

---

Radar is an acronym for *RAdio Detection And Ranging*. Before we develop the principles of radar, we will review the characteristics of radio waves.

1.1 RADIO WAVES

Radio waves occupy a portion of the electromagnetic spectrum from frequencies of a few kilohertz (that is, a few thousand cycles per second) to a few million megahertz ($>10^{12}$ cycles per second). The total electromagnetic spectrum embraces all the frequencies to cosmic rays, beyond 10^{16} megahertz (MHz). Radio waves represent less than one-billionth of the total spectrum (see Appendix 2).

Although electromagnetic energy can be described either as waves or as quanta, the lower frequencies are much better suited to explanations by wave theory. Radio waves are certainly thought of in those terms.

The definitive experiments in electromagnetism were performed by Michael Faraday in a period of ten days in 1831 [*Encyclopedia Britannica,* 1984, Vol. 7,

p. 174; Williams, 1971, pp. 535–537]. Using Faraday's work as his foundation, James Clerk Maxwell succeeded, by the early 1860s, in synthesizing the properties of electricity and magnetism into a set of equations that achieved a unified theory for electromagnetics [*Encyclopedia Britannica,* 1984, Vol. 11, p. 718; Everitt, 1974, pp. 204–217].

In Maxwell's time, it was only dimly appreciated that light is electromagnetic energy and that all electromagnetic energy propagates with the same velocity in free space. Yet, Maxwell's equations, solved for the speed of light, give the correct result. Maxwell's equations are the foundation for the theory and design of modern radio and radar systems. Faraday and Maxwell noted that time-varying electric currents produced time-varying electric and magnetic fields in free space, that these fields would induce time-varying electric currents in materials they encountered, and that these currents would, in turn, generate electric and magnetic fields of their own. These fields "propagate" in free space at the speed of light.

In 1886, Heinrich Hertz conducted a number of experiments showing that radio waves reflected, refracted, were polarized, interfered with each other, and traveled at high velocity. Hertz is credited with verifying Maxwell's theories [*Encyclopedia Britannica,* 1984, Vol. 6, pp. 647–648; McCormach, 1972, pp. 341–350]. These characteristics—of reradiation and of known velocity—already portended the invention of radar.

The first use of radio waves was for communication, and the means of generating radio waves was with spark gaps generating short, intense pulses of current to achieve the needed electromagnetic radiation. The generation of sinusoidal waves (arising first from the use of alternators and later from oscillators designed using the vacuum tubes invented by Lee DeForest in 1906) revolutionized communications [Susskind, 1971, pp. 6–7]. When radar was invented, the use of sinusoidal oscillators was adopted from communications, but the transmitters sent periodic bursts of these sinusoidal waves, carefully counting time between them. More complex modulation schemes for radar came later.

The increasing use of radio in the early 1900s led to observations that objects passing between the transmitters and receivers produced interference patterns (exactly as aircraft today affect television reception). Bistatic (noncollocated transmitter and receiver) CW "radars" that could detect targets in this manner were explored by many countries at the beginning of the 1930s. Monostatic (collocated transmitter and receiver) radars were developed shortly after. The first successful pulsed radar experiments were conducted by the U.S. Naval Research Laboratory (NRL) in 1934. By 1937, NRL had demonstrated a radar at sea, but deployment was delayed until 1940. In the meantime, Great Britain, which earlier had trailed in radar development, succeeded in deploying the first operational system (the Chain Home radars) by 1938. Concurrently, France, Germany, and the Soviet Union also had substantial radar programs underway [Skolnik, 2001, pp. 14–19].

1.2   A SIMPLE RADAR

The principles of a primitive radar are now clear, transmission, propagation, and reflection. A functional diagram of a radar system is shown in Figure 1.1. A pulse of electromagnetic energy, oscillating at a predetermined frequency, f_o, and duration, τ, is generated by the transmitter. The pulse is routed through a transmit-receive switch to an antenna. The transmit-receive switch, or *duplexer*, protects the sensitive receiver from the high-power transmitted pulse. The pulse is radiated into free space through an antenna. The electromagnetic pulse propagates outward at the speed of light, scattering (reradiating) from objects it

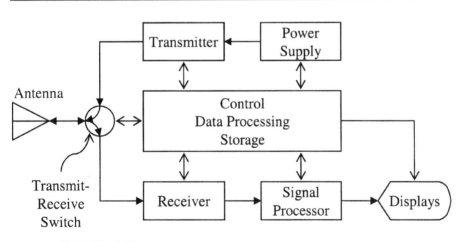

Figure 1.1   Radar block diagram.

encounters along the way. Part of the scattered signal returns to the radar. The scattered signal is collected by the antenna and routed through the transmit-receive switch to the receiver. The presence of the received signal can be detected in the receiver, because it imitates the frequency and the duration of the transmitted pulse. The received signal is enhanced, interfering signals are reduced, and measurements of the object are made by signal processing. The resulting detections of received signals are presented to radar operators on displays.

Detecting the presence of an object is good, but the real value of radar is being able to measure the range between the radar and the object. When a pulse is transmitted, a clock is started. When a received signal is detected, the clock is stopped. Using basic physics—distance equals speed × time—the range to a detected object can be calculated as in Equation (1.1). The distance traveled by the pulse is twice the range—once to the object and once back to the radar.

$$2R = c\,\Delta t \Rightarrow R = \frac{c\,\Delta t}{2} \tag{1.1}$$

where:
   R = range from the radar to the object, meters
   c = speed of light, 3×10^8 meters/second
   Δt = elapsed time, seconds

The transmitted pulsed radar waveform is shown in Figure 1.2. Some other parameters of this basic system for radio detection and ranging are immediately available. The wavelength of the propagated energy is given in Equation (1.2). The angular frequency, ω (radians/sec), is $2\pi f_c$. The time between radar pulses is the interpulse period, or the pulse repetition interval (PRI), and the number of pulses sent per time interval is the pulse repetition frequency (PRF). PRI and PRF are reciprocal, i.e., PRF = 1/PRI. The ratio of the time the pulse is on to the PRI is called the duty cycle [Equation (1.3)]. The average power over one pulse repetition interval is the product of the peak power and duty cycle [Equation (1.4)]. The energy in one pulse is the product of its peak power and duration [Equation (1.5)]. All the previously mentioned transmitted waveform parameters are important in radar design and performance. Most will be discussed in various applications. The impact of these waveform parameters on radar measurements will be discussed in Chapter 5.

$$\lambda = \frac{c}{f_c} \tag{1.2}$$

where:

λ = wavelength, meters

c = speed of light, 3×10^8 meters/second

f_c = radar carrier frequency, hertz

$$d_t = \frac{\tau}{PRI} = \tau\, PRF \tag{1.3}$$

where:

d_t = duty cycle, no units

τ = pulse duration, or pulse width, or pulse length, seconds

PRI = pulse repetition interval, seconds

PRF = pulse repetition frequency, hertz

$$P_{ave} = P\, d_t = P\frac{\tau}{PRI} = P\tau\, PRF \tag{1.4}$$

where:

P_{ave} = average transmit power, watts

P = peak transmit power, watts

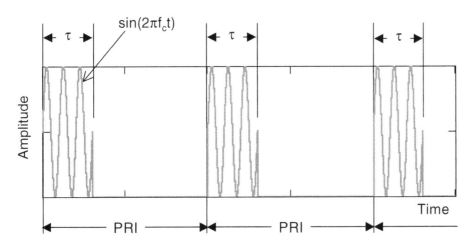

Figure 1.2 Transmitted radar waveform.

$$E = P\tau \qquad (1.5)$$

where:

E = energy in one pulse, watt-seconds or joules

Polarized radar waves are equivalent to polarized light waves [Sears, 1948, pp. 167–185], and the singly polarized radar wave is analogous to polarized optical glasses. Polarization is the orientation of the electromagnetic wave relative to the direction of its propagation. An electromagnetic wave consists of two perpendicular components, the electric field (E-field) and the magnetic field (H-field), as shown in Figure 1.3. Polarization is defined by the alignment of the E-field. Linear polarization describes a linear alignment of the E-field, usually ei-

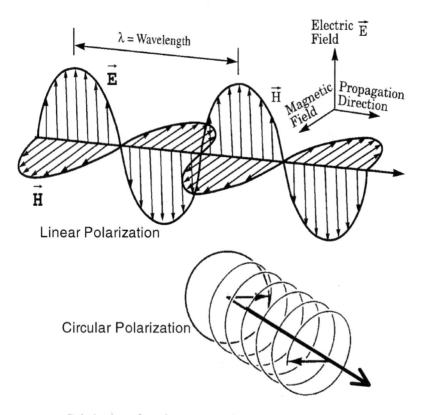

Figure 1.3   Polarization of an electromagnetic wave.

ther horizontal or vertical (as shown in Figure 1.3). Circular polarization describes a rotating vector (one full revolution each radio frequency cycle), either left hand or right hand.

1.3 THE RADAR RANGE EQUATION

The fundamental determinant of radar performance, in any of the missions prescribed for it, is the radar range equation. It can be derived from fundamental principles, as shown in Figure 1.4. Imagine an isotropic source of an electromagnetic pulse of peak power, P, radiating into free space. Provide the source with some focusing device, an antenna that concentrates the power from isotropic to a confined solid angle. Call the ratio of this focusing over isotropic radiation the gain of the transmit antenna, G_T. Antennas will be discussed in detail in Chapter 2. We often represent antenna gain in decibels. Decibels are discussed in detail in Appendix 1. The power density received at range R from the radar is given in Equation (1.6).

$$\frac{PG_T}{4\pi R^2} \tag{1.6}$$

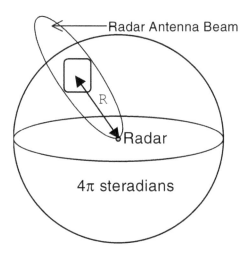

Figure 1.4 Radar spherical geometry.

where:
  P = peak transmit power, watts
  G_T = transmit antenna gain, no units
  R = radar to object range, meters

The power density is intercepted by a target. A portion is reradiated back to the radar, based on the radar cross section, σ, of the target. Radar cross section (RCS) will be discussed in detail in Chapter 4. The power reradiated from the target back to the radar is given in Equation (1.7).

$$\frac{P\,G_T\,\sigma}{4\pi R^2} \qquad (1.7)$$

where:
  σ = target radar cross section, square meters (m^2)

The power density arriving back at the radar from the target is given in Equation (1.8), assuming the transmit and receive antenna are collocated (monostatic). This power density is collected by the effective area of the receive antenna. The received power, now defined as the received signal, S, is given in Equation (1.9). We often represent the received signal in decibels relative to one watt (dBW) or relative to one milliwatt (dBm) (see Appendix 1).

$$\frac{P\,G_T\,\sigma}{(4\pi)^2 R^4} \qquad (1.8)$$

$$S = \frac{P\,G_T\,\sigma\,A_e}{(4\pi)^2 R^4} \qquad (1.9)$$

where:
  S = received signal power, watts
  A_e = effective area of the receive antenna, meters $A_e = \rho\,A$
  ρ = antenna efficiency, $0 < \rho < 1$, no units
  A = physical area of the antenna, meters

The gain of an antenna is directly related to its effective area. In Chapter 2, we will derive the relationship given in Equation (1.10).

$$G = \frac{4\pi A_e}{\lambda^2} \Rightarrow A_e = \frac{G\lambda^2}{4\pi} \tag{1.10}$$

where:

  G = radar antenna gain, no units

If the radar's transmit and receive antenna are the same, the resulting single-pulse received signal power is given in Equation (1.11).

$$S = \frac{P\,G^2\,\lambda^2\sigma}{(4\pi)^3 R^4} \tag{1.11}$$

1.3.1 Receiver Noise

Unfortunately, there is always noise power contaminating the signal power that arrives at the receiver. Some of the noise is generated in the transmitter, some of it is added by the cosmos (galactic noise), some of it is contributed by the Earth's atmosphere (spherics), some is added by the Earth itself, and some from manufactured sources (automobiles, power facilities, or other radars). However, for most radar systems, the vast majority of the noise is generated in the front end of the radar receiver, particularly by the first amplifier and mixer stages. The source of the noise generated in the front end of the radar receiver is the thermal heating of its electronic components. Basic chemistry tells us that when atoms are heated, their electrons flow. Flowing electrons produce current flow, resulting in noise. This is better known as thermal noise, as given in Equation (1.12). We often represent the receiver thermal noise in decibels relative to one watt (dBW) or relative to one milliwatt (dBm). This total system noise can be measured at the receiver output in the absence of signal. The noise figure relates theory to what we can practically achieve. We often represent the noise figure in decibels.

$$N = (F_n - 1)kT_0 B \tag{1.12}$$

where:

  N = receiver thermal noise, watts

  F_n = receiver noise figure, no units

k = Boltzmann's constant, 1.38×10^{-23} watt-seconds/kelvins
T_0 = receiver temperature (usually room temperature, 290 K), kelvins
B = receiver filter bandwidth, hertz

The effective noise temperature is another common way of relating receiver thermal noise theory to what can be practically achieved [Equation (1.13)]. The effective noise temperature is essentially the temperature the receiver would have to reach to produce the resultant noise power.

$$T_e = (F_n - 1) T_0 \qquad (1.13)$$

where:
T_e = effective noise temperature, kelvins

1.3.2 Signal-to-Noise Ratio

The radar range equation is the ratio of the received target signal power to the receiver noise. If the various losses that exist in the system are lumped together in a single term, the radar range equation is given in Equation (1.14). We often represent the signal-to-noise ratio *(S/N)* in decibels. The same goes for the radar system losses.

$$S/N = \frac{P\,G^2\,\lambda^2\,\sigma}{(4\pi)^3 R^4 (F_n - 1)\,k\,T_0\,B\,L_s} \qquad (1.14)$$

where:
S/N = single pulse signal-to-noise ratio, no units
L_s = radar system losses, no units

The radar range equation is dominated by the R^4 factor in the denominator. There is no magic way to achieve a high-performance system. If low RCS targets are to be engaged, a combination of high power, high antenna gain, and low noise seems to be dictated. Fortunately, if needed, and it most often is, we can "integrate" multiple return pulses to improve the *S/N*. Considerable *S/N* benefits can result from integration of multiple pulses [Brookner, 1977, pp. 81–99]. We will discuss integration in more detail in Chapter 3.

1.3.3 Detection Range

The single pulse radar range equation can be solved for the range at which the radar will detect the presence of a target with a given S/N. The radar detection range is given in Equation (1.15). In Chapter 3, we will discuss how to determine the S/N required for detection.

$$R_d = \sqrt[4]{\frac{P\,G^2\,\lambda^2\,\sigma}{(4\pi)^3\,SNR_d\,(F_n-1)\,k\,T_0\,B\,L_s}} \qquad (1.15)$$

where:

R_d = radar detection range, meters

SNR_d = single pulse signal-to-noise ratio required for detection, no units

1.3.4 Other Forms of the Radar Range Equation

There are, however, several ways of manipulating the equation to illustrate various uses. Checking the units of the range equation shows it to be dimensionless. It is pulse power divided by noise power. To emphasize the importance of energy and of matched filters that yield the maximum theoretical signal-to-noise ratio, S/N is often replaced by E/N_0, the maximum ratio of root-mean-square (RMS) signal energy to RMS noise energy [Equation (1.16)]. We will discuss matched filters in more detail in Chapter 3; for now, we will just state that the matched filter bandwidth is the reciprocal of the pulse width. Furthermore, because it facilitates calculations, $2E/N_0$ is also used. The RMS voltage of a sinusoid is equal to the peak voltage divided by the square root of 2, making $2E/N_0$ the peak signal energy divided by the RMS noise energy.

$$\frac{E}{N_0} = \frac{P\,\tau\,G^2\,\lambda^2\,\sigma}{(4\pi)^3\,R^4\,(F_n-1)\,k\,T_0\,L_s} \qquad (1.16)$$

where:

E/N_0 = single pulse energy signal-to-noise ratio, no units

E = RMS signal energy in one pulse, watts/hertz or watt-seconds or joules

N_0 = RMS noise energy, watts/hertz or watt-seconds or joules

Obviously, the range equation can be solved for any of its parameters. Often, it is solved for R, necessitating taking the fourth root of the whole right side of the equation. While this practice is rigorous, it destroys the perspective provided by having those variables in the numerator that multiply signal and those in the denominator that divide it. Boltzmann's constant, $(4\pi)^3$, and conversion factors of meters into kilometers are often accumulated as a single number, easing calculations but removing insights. Examples of this loss of insight are the average power, energy, and surveillance versions of the radar range equation. Each one contains one or more assumptions that are too easily overlooked.

1.4 SURVEILLANCE

Surveillance radars are designed to scan a solid angle in some time span, and the scan program places the necessary number of pulses into each angle bin. Surveillance radars are often required to detect large numbers of targets (aircraft, ships, missiles, and so on) flying around them. Detection must occur before these targets reach a specified minimum range. The surveillance radar designer will select the frequency of the radar at which the combination of the physical size of the radar aperture and the generation of radar power are least expensive. Because both characteristics favor the use of lower microwave frequencies, the vast majority of surveillance radars are at those frequencies. A survey of such radars around the world, including both U.S. and Russian missile and aircraft warning and surveillance systems, confirms the truism. Virtually all these radars are in the frequency band between 200 and 1500 MHz (VHF to L-band).

Big apertures and high-powered transmitters are characteristic of surveillance radars. Average power of a few hundred kilowatts (kW) (with from 10 to 50 megawatts [MW] of peak power, depending on duty cycle) is possible for these radars. For reflector antenna radars, such power can be generated by several oil-cooled klystrons operated in parallel. For phased array radars, transistor amplifiers may be used behind each element (100 W each behind several thousand elements) or large tubes (traveling wave tubes or klystrons) behind groups of elements.

Apertures may range in diameter from 1 to over 120 m for both reflector antennas and arrays, giving gains of about 40 dB and beamwidths on the order of 1 square degree.

1.5 TRACKING

Tracking radars have different characteristics from those of surveillance radars. We discuss tracking in Chapter 3. Here, we will limit comments to saying that the driving parameter in good tracking is short wavelength. Not surprisingly, tracking radars (including the ground-controlled approach radars at airports and missile trackers of the military) are at the high microwave frequencies (from 3 to 20 GHz). For the designer of a radar requiring a combination mission of both surveillance and tracking, the relative desirability of high frequency for trackers and big apertures for surveyors poses a dilemma. The usual procedure is to write down a cost function and settle the matter of frequency by optimizing costs with respect to wavelength. Specific examples of how many existing radars obey the dictates of the previously listed functions are these: The Lincoln Laboratory ALCOR tracking radar at Kwajalein Atoll in the South Pacific is at C-band ($\lambda \cong 0.08$ m); the Ballistic Missile Early Warning and PAVE PAWS submarine launched ballistic missile (SLBM) warning radars in Canada, Greenland, Scotland, Massachusetts, and California are at UHF, $\lambda \cong 0.6$ m.

1.6 EXERCISES

1. The Anti-Ballistic Missile Defense Treaty of the Strategic Arms Limitations accords with Russia limits radars at ABM sites in the two countries to 3×10^6 watt-meters-squared of power-aperture product. The radar designers want to use a 250-kW peak power transmitter, operating at a frequency $f_c = 6$ GHz. Assuming that such a radar requires a signal-to-noise ratio $S/N = 10$ (10 dB) for detection, a bandwidth $B = 300$ kHz, effective noise temperature $T_e = 1000$ K, and system losses $L_s = 10$ (10 dB), what would be its maximum possible detection range against a 10-m^2 target?

2. If a radar with transmit peak power $P = 1$ MW peak power and antenna gain $G = 1000$ (30 dB) irradiates a target with an RCS $\sigma = 1$ m^2 target at a range $R = 500$ km range, what power density arrives back at the radar antenna?

3. The radar in Exercise 2 is transmitting at $f_c = 1$ GHz. If the receive side of this radar has a noise figure $F_n = 3$ dB, receiver bandwidth $B = 100$ kHz, and system losses $L_s = 10$ (10 dB), what receiver temperature, T_0 (kelvins) is required to give the single pulse signal-to-noise ratio of unity?

4. The cost of a radar is the cost of power plus the cost of aperture plus a constant. The cost of power is the cost/kilowatt multiplied by the number of

kilowatts: $C_p = C_{kW} N_{kW}$. In addition, the cost of aperture is the cost per square meter multiplied by the number of square meters: $C_A = C_m^2 N_m^2$. Show that, for minimum cost,

$$2 C_p = C_A$$

(Hint: Assign a design power-aperture product.)

5. What is the range associated with a time delay $\Delta t = 0.67$ msec? What is the time delay associated with the range to the moon (Earth-to-moon range $R = 3.84 \times 10^8$ m)?

6. What is the antenna gain for an antenna with the following characteristics: 5 m width, 2 m height, efficiency $\rho = 50\%$, and frequency $f_c = 3$ GHz?

7. What is the receiver thermal noise for the following receiver characteristics: noise figure $F_n = 6$ dB, standard temperature $T_0 = 290$ K (room temperature), and receiver bandwidth B = 750 kHz?

1.7 REFERENCES

Brookner, E. (Ed.), 1977, *Radar Technology*, Norwood, MA: Artech House.

Encyclopedia Britannica, 15th ed., 1984, Chicago: Benton. Maxwell and Faraday were Englishmen, *Britannica* does well by them.

Everitt, C. W. F., 1971, James Clerk Maxwell, in C. C. Gillespie (Ed.), *Dictionary of Scientific Biography*, New York: Scribner's.

McCormach, R., 1972, Heinrich Hertz, in C. C. Gillespie (Ed.), *Dictionary of Scientific Biography*, New York: Scribner's.

Sears, F. W., 1948, *Principles of Physics*, Volume 3, Optics, Cambridge, MA: Addison-Wesley.

Skolnik, M., 1985, "Fifty Years of Radar," *Proceedings of the IEEE*, Feb. 1985.

Susskind, C., 1971, Lee DeForest, in C. C. Gillespie (Ed.), *Dictionary of Scientific Biography*, New York: Scribner's.

Watson-Watt, R., 1957, *Three Steps to Victory*, London: Odham's Press.

Watson-Watt, R., 1959, *The Pulse of Radar*, New York: Dial Press. Watson-Watt's books are both written in popular style but with important technical issues discussed.

Williams, L. P., 1971, Michael Faraday, in C. C. Gillespie (Ed.), *Dictionary of Scientific Biography*, New York: Scribner's. The *Dictionary of Scientific Biography* provides biographical information, briefly and in a narrative style, but also includes technical contributions and some mathematical developments.

2

Antennas

HIGHLIGHTS

• Antenna gain and effective area

• Remarkable utility of the paraboloid

• Deriving the antenna far-field antenna gain pattern with calculus

• Design features for mainbeams and sidelobes

• Unique features of arrays

• Several ways to steer the beams of phased arrays, emphasizing phase shifters

An antenna is the mechanism by which the electromagnetic signal is radiated and received. For radars (although not necessarily for antennas in other electromagnetic applications), it is essential that the antenna enhance performance. A radar antenna has three roles: to be a major contributor to the radar's detection performance, to provide the required surveillance, and to allow measurements of angle of sufficient accuracy and precision.

A reasonable place to begin is with the two expressions used in Chapter 1 to derive the radar range equation: *antenna gain* and *effective area*. For a transmitting antenna, antenna gain is simply a measure of how much focusing of the transmitted waveform is being accomplished by the antenna. Focusing is the ability to add up energy preferentially. Energy arriving at the antenna from a given direction is integrated; that arriving from elsewhere is not. It is assumed that energy arriving at the antenna is in the form of plane waves, that is, the phase of the arriving energy is constant over any plane perpendicular to the direction of arrival, as shown in Figure 2.1. Because most sources of electromag-

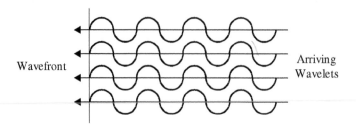

Figure 2.1   A plane wave.

netic energy are small, these wavefronts are really spherical, but at ranges of interest, the approximation to a plane wave is good.

For a receiving antenna, effective area is simply a measure of ability of the antenna to intercept the incident power density. Electromagnetic theory tells us the relationship between antenna gain and the effective area of the antenna [Equation (2.1)]. Effective area is related to the physical area of the antenna by the antenna efficiency. The antenna efficiency term allows us to account for factors such as antenna manufacturing tolerances, illumination function, feed networks, array element characteristics, and so forth.

$$G = \frac{4\pi \, A_e}{\lambda^2} = \frac{4\pi \, \rho \, A}{\lambda^2} \tag{2.1}$$

where:
  G = antenna gain, no units
  A_e = effective area of the antenna, meters
  λ = wavelength, meters
  ρ = antenna efficiency, $0 < \rho < 1$, no units
  A = physical area of the antenna, meters

2.1 A PARABOLIC REFLECTOR

The classical shape for focusing electromagnetic energy is the parabolic reflector (often referred to as a *dish*). Recall that a parabola (depicted in Figure 2.2) has interesting characteristics [Gardner, 1981]. Parallel lines, drawn from a line perpendicular to the axis of the parabola to the parabola and thence to its fo-

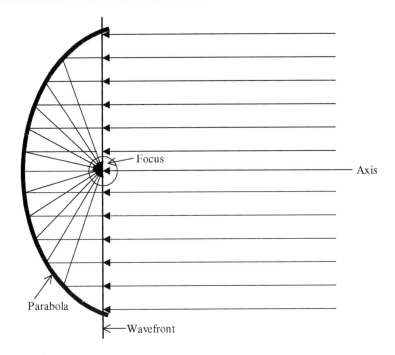

Figure 2.2　A parabolic antenna.

cus, will all be of the same length. Consequently, arriving plane waves will maintain constant phase over this distance. Furthermore, the angle made by the reflected line (with a line tangent to the parabola at that point) is always the same as the angle made by the incident line. If the lines are thought of as rays, then the angle of incidence equals the angle of reflection, which is Snell's law. The application to electromagnetic waves is obvious. All waves arriving parallel to the axis of the parabola and oscillating in phase will add up at the focus; other arriving waves will not add up. In effect, the parabola is focused on infinity. Only waves arriving from a source at infinity (a few miles approximating infinity in practice) will be parallel. The same can be said for radiating waves. Energy leaving the focus and reflecting off the parabola will radiate as a plane wave. When extended to a third dimension, the figure becomes a paraboloid. Put a small antenna, such as a feed horn, at its focus, and we have a parabolic reflector. Mount it on a pedestal that allows it to pivot on two axes, and we have the canonical radar antenna.

Notice that any segment of the parabola has equivalent characteristics. If only the top one-third of the reflector is illuminated, the feed is offset and does not interfere with the incident radiation. Reflectors with offset feeds are commonplace in radar. The feed might also be embedded in the center of the reflector, radiating its energy outward against a plate at the paraboloid's focus, which in turn would reflect it against the reflector. For the geometry to work, the "splash plate" must be a hyperboloid. This arrangement is known as a Cassegrain feed, after the optical telescopes of similar design. Of course, these applications require careful design of the feeds, which are antennas themselves.

Traditionally, reflector antennas have modest antenna efficiencies. Computer-controlled manufacturing has greatly improved the precision of reflector antennas and thus increased their efficiencies.

The plane wave generated by the two-dimensional parabola could also be generated by a line of incremental sources generating sinusoidal waves that are in phase. Being "in phase" means being in the same point of the sinusoidal wave at the same instant in time, as shown in Figure 2.3. we will postulate that these sources radiate semi-isotropically (that is, in a semicircle above a ground plane) and that they are sufficiently close together that they are "indistinguishable" from an extended source of in-phase energy. (How to determine what is "indistinguishable" is described in Chapter 8 under "Far Field of an Antenna," p. 219.) Because there can be a change in these phase relationships as a function of angle, the antenna gain varies with angle as the waves from each source go in and out of phase with respect to each other.

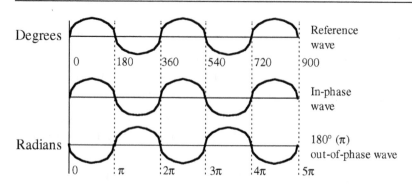

Figure 2.3 Phase relationships.

2.2 THE ANTENNA PATTERN

It is one thing to have found the antenna gain; it is another to know how its value varies across the spectrum of angles—in other words, to know the antenna pattern. There are three main approaches for determining the antenna pattern: electromagnetic theory, Fourier transforms, and an incremental algebraic. We will concentrate on the electromagnetic theory approach and identify how the other approaches can be used.

We will start with a simple line antenna with a current distribution across the length of the antenna, as shown in Figure 2.4. Often, the current distribution is called an *illumination function*. Electromagnetic theory states that an electric field (E-field) is produced perpendicular to the current distribution. The relationship among the current distribution, antenna dimensions, and E-field is given in Equation (2.2). We will solve this equation for the uniform current distribution case (current distribution is constant across the antenna), as shown in Figure 2.5. For a uniform current distribution, Equation (2.2) is rewritten as Equation (2.3). The integral in Equation (2.3) is solved, and evaluated at its limits, as given in Equation (2.4). Equation (2.4) looks pretty formidable until we use the trigonometry identity given in Equation (2.5). Using this trigonometry identity, the E-field is now given in Equation (2.6). Often, we normalize the E-field so $E(0) = 1$; this is accomplished by setting $A_0 = 1/D$. The normalized E-field is given in Equation (2.7). The E-field is the antenna voltage pat-

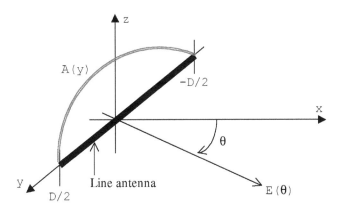

Figure 2.4 Simple line antenna.

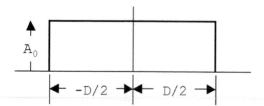

Figure 2.5 A uniform current distribution.

tern. The antenna gain (power) pattern is the square of the E-field, as given in Equation (2.8).

$$E(\theta) = \int_{-D/2}^{D/2} A(y) \, e^{\left(\frac{2\pi \, j \, y \, \sin(\theta)}{\lambda}\right)} dy \qquad (2.2)$$

where:

$E(\theta)$ = E-field produced by the current distribution, perpendicular to $A(y)$, voltage antenna gain pattern, no units
D = antenna length, meters
$A(y)$ = current distribution in the y-z plane
λ = wavelength, meters

$$E(\theta) = \int_{-D/2}^{D/2} A_0 \, e^{\left(\frac{2\pi \, j \, y \, \sin(\theta)}{\lambda}\right)} dy \qquad (2.3)$$

$$E(\theta) = \frac{A_0}{\left(\frac{2\pi \, j \, y \, \sin(\theta)}{\lambda}\right)} \left[e^{\left(\frac{2\pi \, j \, y \, \sin\theta}{\lambda}\right)} \right]_{-D/2}^{D/2}$$

$$E(\theta) = \frac{A_0}{\left(\frac{2\pi \, j \, y \, \sin(\theta)}{\lambda}\right)} \left[e^{\left(2\pi \, j \frac{D}{2} \sin(\theta)\right)} - e^{\left(-2\pi \, j \frac{D}{2} \sin(\theta)\right)} \right]$$

$$\qquad (2.4)$$

$$\sin(\phi) = \frac{e^{j\phi} - e^{-j\phi}}{2j} \tag{2.5}$$

$$E(\theta) = \frac{A_0 D}{\left(\frac{\pi D \sin(\theta)}{\lambda}\right)} \sin\left(\frac{\pi D \sin(\theta)}{\lambda}\right) \tag{2.6}$$

$$E(\theta) = \frac{\sin\left(\frac{\pi D \sin(\theta)}{\lambda}\right)}{\left(\frac{\pi D \sin(\theta)}{\lambda}\right)} \tag{2.7}$$

$$G(\theta) = [E(\theta)]^2 = \left[\frac{\sin\left(\frac{\pi D \sin(\theta)}{\lambda}\right)}{\left(\frac{\pi D \sin(\theta)}{\lambda}\right)}\right]^2 \tag{2.8}$$

where:

$G(\theta)$ = antenna gain pattern (power) for a uniform current distribution, no units

The antenna gain pattern can be plotted as an x-y plot or a polar plot, as shown in Figure 2.6. The antenna gain pattern is characterized by its main-beam, sidelobes, nulls, and backlobe, as shown in Figure 2.7. It is commonplace for antenna gain pattern to be plotted in decibels, as in both of these figures. Decibels are used so the actual pattern can be easily visualized. The peak of the mainbeam is the antenna gain, as computed using Equation (2.1). The majority of the antenna gain is concentrated in the mainbeam, with the remaining an-tenna gain distributed in the sidelobes and the backlobe.

2.2.1 Finding Half-Power and Null-to-Null Beamwidths

We commonly use two different measures of the angular width of the main-beam: the half-power beamwidth and the null-to-null beamwidth. The half-power beamwidth is the angular distance between the half-power points of the mainbeam. We start the process of determining the half-power beamwidth by

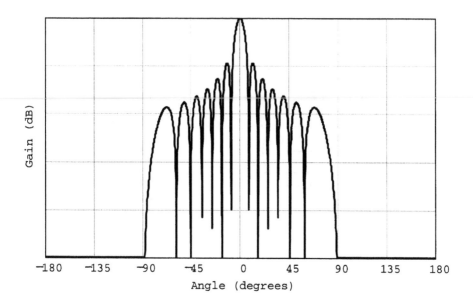

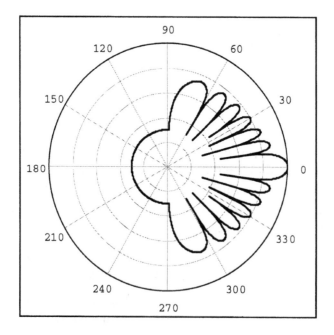

Figure 2.6   Antenna gain pattern, x-y plot and polar plot.

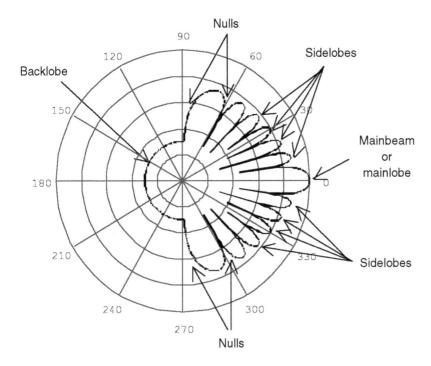

Figure 2.7   Antenna pattern characteristics.

solving for the angle at which Equation (2.8) equals 0.5, as given in Equation (2.9). The $[\sin(x)/x]^2$ function equals 0.5 when $x = 0.44$. We use this relationship in Equation (2.10). Next, we use the small angle approximation for sine $[\sin(\theta) \approx \theta$ for small values of θ in radians], as given in Equation (2.11). We then write the equation for the half-power beamwidth, Equation (2.12). Converting from radians to degrees gives us the half-power beamwidth in Equation (2.13). A common approximation for the beamwidth in radians is λ/D, although we can see by Equation (2.12) how approximate this is.

$$G(\theta) = \left[\frac{\sin\left(\dfrac{\pi D \sin(\theta)}{\lambda}\right)}{\left(\dfrac{\pi D \sin(\theta)}{\lambda}\right)}\right]^2 = 0.5 \qquad (2.9)$$

$$\left(\frac{\pi D \sin(\theta)}{\lambda}\right) = 0.44\pi$$

$$\sin(\theta) = \frac{0.44\lambda}{D} \tag{2.10}$$

$$\theta \cong \frac{0.44\lambda}{D} \tag{2.11}$$

$$\theta_{3dB} = 2\theta = \frac{0.88\lambda}{D} \text{ radians} \tag{2.12}$$

$$\theta_{3dB} = \frac{51\lambda}{D} \text{ degrees} \tag{2.13}$$

where:

θ_{3dB} = antenna half-power (–3 dB) beamwidth for a uniform current distribution, radians or degrees

We can determine the null-to-null beamwidth in a similar manner as the half-power beamwidth. Nulls in the antenna gain pattern occur when the argument of the sine term equals π [$\sin(\pi) = 0$], as given in Equation (2.14). Solving for $\sin(\theta)$, we obtain Equation (2.15). Once again, we use the small angle approximation for sine [$\sin(\theta) \approx \theta$ for small values of θ in radians], as given in Equation (2.16). We can then write the equation for the null-to-null beamwidth, Equation (2.17). Converting from radians to degrees gives us the null-to-null beamwidth in Equation (2.18).

$$\left(\frac{\pi D \sin(\theta)}{\lambda}\right) = \pi \tag{2.14}$$

$$\sin(\theta) = \frac{\lambda}{D} \tag{2.15}$$

$$\theta \cong \frac{\lambda}{D} \tag{2.16}$$

$$\theta_{nn} = 2\theta = \frac{2\lambda}{D} \text{ radians} \tag{2.17}$$

$$\theta_{nn} = \frac{115\lambda}{D} \text{ degrees} \tag{2.18}$$

where:

θ_{nn} = antenna null-to-null beamwidth for a uniform current distribution, radians or degrees

2.2.2 Finding Sidelobe Levels

We have seen that the antenna gain pattern is of the form $[\sin(x)/x]^2$. We can now manipulate the integral of this function, as given in Equation (2.19), to find out several interesting things. By computing the ratio of an integration from 0 to 0.44π (the half-power point) and from 0 to ∞, we can find that just under 50% of the antenna gain is contained within the half-power beamwidth. By computing the ratio of an integration from 0 to π (the first null) and from 0 to ∞, we can find just under 90% of the antenna gain is contained within the null-to-null beamwidth.

$$\int_0^x \left(\frac{\sin x}{x}\right)^2 dx \tag{2.19}$$

We have already noted that the nulls in the antenna pattern are at λ/D, $2\lambda/D$, $3\lambda/D$, and so forth. Peaks in the antenna gain pattern occur when the argument of the sine term equals 0, $3\lambda/2D$, $5\lambda/2D$, and so forth. (They are very nearly so.) The first peak is the antenna mainbeam, or mainlobe, occurs at zero. The second peak, the peak of the first sidelobe, occurs at $3\lambda/2D$. The antenna gain at the peak of the first sidelobe is computed as given in Equation (2.20). This means the antenna gain of the peak of the first sidelobe of an antenna with a uniform current distribution is down from the mainbeam by a factor of 22.2, or 13.46 dB. The antenna gain of the peak of the nth sidelobe of an antenna with a uniform current distribution can be readily defined by Equation (2.21). The tenth sidelobe, therefore, would be 1/1088, or −30.37 dB.

$$\left(\frac{\sin\left(\frac{3\pi}{2}\right)}{\frac{3\pi}{2}}\right)^2 = \frac{1}{22.2} = 0.045 \tag{2.20}$$

$$\left(\frac{2}{\pi}\right)^2 \frac{1}{(2n+1)^2} \tag{2.21}$$

By using a different current distribution, different relationships between an-tenna beamwidth and sidelobe levels can be obtained. Generally, lower side-lobe levels are desired and thus, current distributions with less amplitude further out in the distribution (toward the edge of the antenna) are used. One current distribution that appears naturally is the cosine function, as shown in Figure 2.8. The feedhorns used to radiate energy into a parabolic reflector have antenna patterns shaped like a cosine in amplitude (cosine^2 in power). Dipoles also have a similar far field pattern. Thus, most reflector antennas have current distributions with cosine weighting. What does this do to the far field pattern of the reflector antenna? We can find out by solving Equation (2.2) when a cosine current distribution is used, as given in Equation (2.22), and then squaring the result to determine the antenna gain pattern.

$$E(\theta) = \int_{-D/2}^{D/2} \cos(y) e^{\left(\frac{2\pi jy \sin(\theta)}{\lambda}\right)} dy \tag{2.22}$$

Solving this integral is left to the reader (see Exercise 3 at the end of this chapter). The antenna gain pattern for an antenna with a cosine current distri-bution is shown in Figure 2.9. The peak of the first sidelobe is 23 dB down from

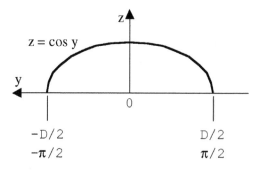

Figure 2.8 A cosine current distribution.

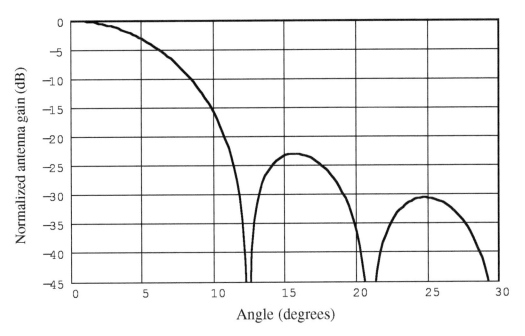

Figure 2.9 Far field of a cosine current distribution (one side).

the mainbeam gain. The lower sidelobe levels come at a price; the mainlobe is 37% wider than for the uniform current distribution.

In trying to select a near-optimum illumination function, the radar designer must do extensive trade-off analyses, particularly on the transmit side, because any weighting of the illumination function from uniform vitiates the power density (watts/m^2) in the mainbeam by a substantive amount. Unfortunately, all sidelobe suppression techniques fatten the main beam and reduce its peak level. Table 2.1 shows representative illuminations functions with their mainbeam performance, both beamwidth and relative gain, and the first sidelobe level below peak. The beamwidth increase and gain loss penalties would have to be made up elsewhere in the system. Many illumination functions have been used for radar antennas other than those listed in Table 2.1. Two of the more common illumination functions are Dolph-Chebyshev and Taylor. The main characteristic of these illumination functions is the ability to design to a specific sidelobe level. For a Dolph-Chebyshev illumination function, the peak sidelobe

Table 2.1 **Characteristics of Different Illumination Functions**

Illumination Function	Beamwidth (deg)	Gain Loss	First Sidelobe Level (dB)
Uniform: $A(y) = 1$	$51\ \lambda/D$	1	−13.2
Cosine: $A(y) = \cos^n\left(\dfrac{\pi}{2}\,y\right)$			
$n = 0$	$51\ \lambda/D$	1	−13.2
$n = 1$	$69\ \lambda/D$	0.810	−23
$n = 2$	$83\ \lambda/D$	0.667	−32
$n = 3$	$95\ \lambda/D$	0.575	−40
$n = 4$	$111\ \lambda/D$	0.515	−48
Parabolic: $A(y) = 1 - (1 - \Delta)\,y^2$			
$\Delta = 1$	$51\ \lambda/D$	1	−13.2
$\Delta = 0.8$	$53\ \lambda/D$	0.994	−15.8
$\Delta = 0.5$	$56\ \lambda/D$	0.970	−17.1
$\Delta = 0$	$66\ \lambda/D$	0.833	−20.6
Triangular: $A(y) = 1 - |y|$	$73\ \lambda/D$	0.75	−26.4
Circular: $A(y) = \sqrt{1 - y^2}$	$58.5\ \lambda/D$	0.865	−17.6
Cosine^2 on a pedestal:			
$0.33 + 0.66\cos^2\left(\dfrac{\pi}{2}\,y\right)$	$63\ \lambda/D$	0.88	−25.7
$0.08 + 0.92\cos^2\left(\dfrac{\pi}{2}\,y\right)$, Hamming	$76.5\lambda/D$	0.74	−42.8

levels are all the same, just like a Chebyshev filter response. For a Taylor illumination function, the peak sidelobe levels are the same for a select number of close in sidelobes (e.g., the first three or four). The relationships among beamwidth, gain loss, and first sidelobe levels for a Dolph-Chebyshev or Taylor are more complex than those given in Table 2.1.

Figure 2.10 shows the sidelobe behavior of a few of the illumination functions of Table 2.1. Note that the mainbeams of the weighted illumination functions extend into the first sidelobe of the uniform illumination function. Thus, quoting the level of the first sidelobe of various illumination functions alone is misleading; the increase in beamwidth must also be given.

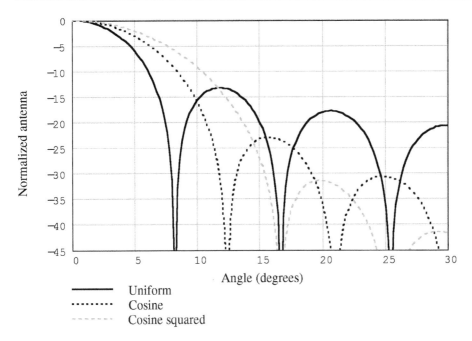

Figure 2.10 Antenna patterns for some illumination functions.

2.2.3 Finding the Antenna Gain within the Antenna Gain Pattern

Useful sidelobe level approximations for radar antennas without detailed calculations have long been sought. The only one that is accurate is that, when integrated over 4π steradians, the gain of an antenna must be 1. One approximation is to define an average sidelobe level for the entire sidelobe region. As we have seen, there are many ways to make sidelobes arbitrarily low by weighting the illumination functions of the antenna; thus, there is no simple way to determine the average sidelobe other than "eyeballing" the antenna gain pattern. However, this simple relationship can give insight. If one needs to find the antenna gain at a specific angle in the antenna gain pattern, a more detailed and accurate approach is used. All one needs is the antenna gain pattern and the mainbeam gain. For example, if the antenna in Figure 2.9 has a mainbeam gain of 32 dB, the antenna gain at an angle of 10° is 32 dB – 15 dB = 17 dB. Likewise, the antenna gain at an angle of 20° is 32 dB – 35 dB = –3 dB.

Sometimes an approximation will do, but most times we need an accurate antenna gain value at a specific angle.

2.3 ARRAY RADARS

An important class of radars uses arrays instead of reflectors for its antennas [Allen, 1963; Hansen, 1966; Stark, 1974; Skolnik, 2001; Stimson, 1998]. Array antennas are the distant past, present, and future of radar systems. Some illumination functions that are difficult or impossible to achieve with reflector antennas can be straightforwardly achieved with an array. Array antennas can also electronically steer the antenna beam, so the time and mechanical stresses of moving the antenna beam around the sky are eliminated. An electronically steerable array, whose beam steering is essentially inertialess, is more complex and capable of less precision than the reflector antenna, but it is much more cost effective when the mission requires surveying large solid angles while tracking large numbers of targets and perhaps guiding interceptors as well. Whereas one or more reflector antennas might handle tens of targets simultaneously, if hundreds, or even thousands, of targets are involved, electronic steering is the only practical answer. That is why missions such as space-track, submarine-launched missile warning, multiple target track/engage fire control, and navy battle group defense all use array antennas in various forms.

The big ballistic missile early warning system (BMEWS) array radar in Thule, Greenland, is used for warning of the launch of intercontinental ballistic missiles. The space-track array radar in Florida is used to catalog space objects. The Cobra Dane array radar in Shemya, Alaska, gathers intelligence data. The Navy's Aegis defense system uses array radars, as do some modern airport approach control radar systems.

In somewhat less demanding missions, electronic scanning may be used in one dimension while mechanical scanning is used in the other. Some air traffic control radars have this feature, as do some air-to-air and air-to-ground airborne radars. Generally, these radars electronically scan in the elevation plane and mechanically scan in azimuth.

Although there are many approaches to achieving electronic beam steering, it must be done by time-delay networks, phasing networks, or a combination of the two. Time-delay steering is the simplest to grasp conceptually. A beam can be formed in almost any direction by adjusting the time at which energy is permitted to emerge from each differential source, or element. In Figure 2.11, a wavefront is formed at an arbitrary angle by delaying by progressive time increments the emission of energy across segments of the one-dimensional array. A beam is formed pointing in the desired angular direction by the relationship between elements given in Equation (2.23).

$$\theta_0 = \sin^{-1}\left(\frac{c\,\Delta t}{d}\right) \tag{2.23}$$

where:
θ_0 = desired mainbeam pointing angle relative to the array boresight, radians, or degrees
c = speed of light, 3×10^8 meters/second
Δt = time delay between successive array elements, seconds
d = array element spacing, meters

True time-delay arrays have been built, but consider the complexity: for n beam positions, we have $n\,\Delta t$ time-delay networks at each element. For an N-

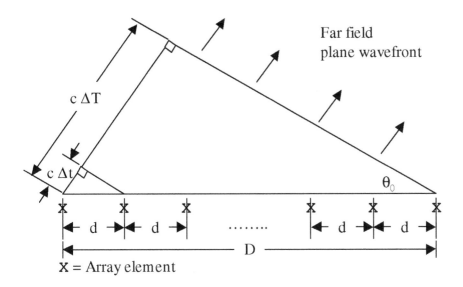

Figure 2.11 Time-delay steering.

element array, we have $N \times n \times \Delta t$ time-delay networks. Because, for practical radar systems, n can be hundreds and N thousands, the result is $>10^5$ time delays. Within any radar pulse, however, only the phase relationships count; so, provided the pulse is long compared to the size of the array (that is, $c\tau \gg D$), time-delay networks need not be used, and the phase relationship across the elements determines the beam direction. From Figure 2.12, assuming sinusoidal signals, the ith element produces a signal $V_i \cos(\omega t + \beta_i)$. A beam is formed pointing in the desired angular direction by the phase relationship given in Equation (2.24). The entire phase progression across the array is given in Equation (2.25).

$$\beta_i = \frac{2\pi\, d_i}{\lambda} \sin\theta_0 \tag{2.24}$$

$$B = \sum_{i=1}^{N} \beta_i$$

$$B = \frac{2\pi\, D}{\lambda} \sin\theta_0 \tag{2.25}$$

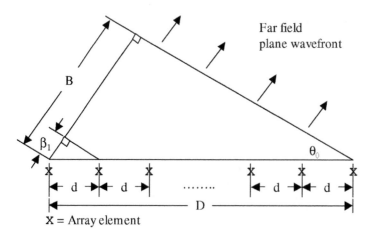

X = Array element

Figure 2.12 A phase-steered array.

where:

β_i = phase shift at the ith element, radians

d_i = distance of the ith array element from the start of the array, meters

λ = wavelength, meters

B = entire phase progression across the array, radians

N = number of array elements, no units

D = length of the array, meters

θ_0 = desired mainbeam pointing angle relative to the array boresight, radians

Observe that an equivalent β_i may be achieved for any θ_0 by making λ a variable. The array can thus be made to point in an arbitrary direction by changing the frequency. To keep the change in λ to a usable level, say ±5%, it is necessary to have the electrical distance between elements greater than 5λ. The ratio of this electrical distance to the physical distance is called the *wrap-up factor*. Some electronic beam steering is done in just this way and is called *frequency scanning*. For frequency-scan arrays, the direction the mainbeam points is given in Equation (2.26). If the mainbeam is to be steered over the angular limits ±θ_1, the necessary wavelength (or frequency excursion) is given in Equation (2.27). Several U.S. and Russian radars use this type of beam steering. Its advantage of relative simplicity must be weighed against its disadvantage of extremely narrow signal bandwidth (a wideband signal causes steering of the beam). Consequently, several other ways to steer beams electronically have been found.

$$\sin(\theta_0) = \frac{L}{d}\left(1 - \frac{\lambda}{\lambda_0}\right) = \frac{L}{d}\left(1 - \frac{f_0}{f}\right) \tag{2.26}$$

$$\sin(\theta_1) = \frac{L}{2}\frac{\Delta\lambda}{d}\cong\frac{L}{2d}\frac{\Delta f}{f_0} \tag{2.27}$$

where:

θ_0 = beam steering angle, radians or degrees

L = electrical distance between array elements, meters

d = physical distance between array elements, meters

λ = transmitted wavelength, meters

λ_0 = wavelength corresponding to the beam pointing at broadside, meters

f_0 = frequency corresponding to the beam pointing at broadside, hertz

f = transmitted frequency, hertz

θ_1 = beam steering limits, radians or degrees
$\Delta\lambda$ = wavelength excursion, meters
Δf = frequency excursion, hertz

2.3.1   Beam Steering with Phase Shifters

Currently, the most popular way to perform beam steering is with phase shifters. This is why so many electronically scanned array antennas are called *phased arrays*. Diode and ferrite phase shifters are used [Stark, Burns, and Clark, 1970; Temme, 1972; Skolnik, 2001]. Diode shifters are appealing because they are small, relatively inexpensive, inherently digital, and easy to design and to understand. Because there are so many beam positions and so many elements contributing to the beam, mixing a relatively few discrete phases at each element can efficiently steer the beam. Phase shifting can be made economic. N two-pole single throw switches (like a diode, which can be turned either on or off) allow for 2^N different phases. This situation is illustrated in Figure 2.13, which shows a two-bit phase shifting system.

Tabulated below are the four different phase shifts available from the two switches.

Switch Positions		Resulting Phase Shift
1	1	0°
1	2	90°
2	1	180°
2	2	270°

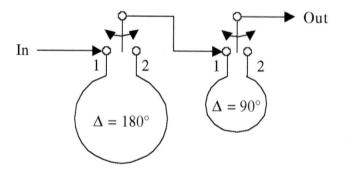

Out

In

1   2   1   2

$\Delta = 90°$

$\Delta = 180°$

Figure 2.13   Concept of a two-bit phase shifter.

The higher the number of different phase shifts, the more precisely the beam can be steered, as given in Equation (2.28). Thus, two-bit phase shifters are too coarse for practical use, but three-, four-, and five-bit shifters are used. With only three, four, and five discrete delay loops, they can generate 8, 16, and 32 phase variations. Phase shifts of more than 2π radians are redundant. (Mathematicians say that phase shift amounts are "modulo 2π"; that is, a phase shift of $2n\pi + \Delta$ is not different from one of Δ [n an integer].) These combinations are used between $0°$ and $360°$ of phase shift. A three-bit phase shifter might have $45°$, $90°$, and $180°$ increments, thereby allowing these combinations:

1. $0°$
2. $45°$
3. $90°$
4. $90° + 45° = 135°$
5. $180°$
6. $180° + 45° = 225°$
7. $180° + 90° = 270°$
8. $180° + 90° + 45° = 315°$

$$\Delta\theta_0 = \frac{\pi}{4}\frac{1}{2^B}\,\theta_{3dB} \tag{2.28}$$

where:

$\Delta\theta_0$ = maximum beam steering error, radians, or degrees
B = number of different phase shifts, no units
θ_{3dB} = array beamwidth, radians or degrees

The same things could be done less efficiently with seven $45°$ phase shifts. A four-bit phase shifter, with 16 available combinations using increments of $22.5°$, can be used, or five bits with $11.25°$. If variable phase shifters are used, the beam can be steered as relative phase between elements is finely changed. Of course, phase shifter beam positioning will not be done exactly as described. For example, randomization at the points of discontinuity along the phase gradient is essential to keep the sidelobes smooth. For big radar antennas, designers tend to think of three-bit phase shifters as not enough and four-bit shifters as too much.

2.3.2   Other Beam Steering Approaches

Huggins beam steering was used in some early phased arrays. An analog concept, Huggins beam steering employed double mixing. The objective is to radiate the carrier radian frequency, $\omega_0 t$, with a phase difference appropriate for each element. To do this requires mixing (multiplying) the carrier frequency with the phasing frequency and filtering out unwanted products.

The so-called Butler array or Butler matrix can be used for signal processing as well as for beam steering [Moody, 1964; Shelton, 1968]. Although its plumbing is complicated, its concept is not, as shown in Figure 2.14. A hybrid is a waveguide junction that is a directional coupler. If energy is put into arm one, half goes to arm 2 in phase and half to arm 3 with 90° advance in phase. No energy goes to arm 4. A signal at one of the inputs produces a beam formed by the phase front produced by radiation from all the outputs. Everything about the Butler array is orthogonal. A four-element input-output matrix forms four beams in space that cross over at the 4-dB points. An interesting property of the Butler matrix is that if the spectral components of a wave are inserted at the outputs, their time function will appear at an input. Because of the analogies previously presented between time-frequency and aperture-angle, this result

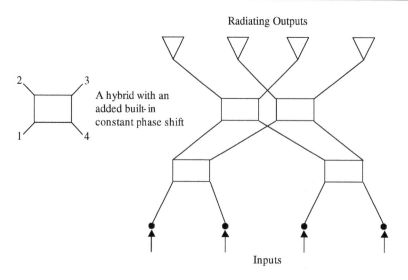

Figure 2.14   A Butler array.

should not be too surprising. The transformation is limited in its granularity. In Figure 2.14, only four range resolution cells can be accommodated. However, the operations are few, and the Butler matrix is often used for signal processing, either actually or implicitly, in a computer. When a Butler matrix is used in this way, its outputs are called "fast Fourier transforms."

The Blass array is another beam-steering technique [Blass, 1960]. It is diagrammed in Figure 2.15. The differences in physical lengths form the phase fronts of the multiple beams. A Blass array can form beams arbitrarily close together in angle space. Thus, it can effectively turn continuous functions of angle into illumination functions or continuous functions of frequency into time functions. In other words, it can perform actual Fourier transforms. A comparison between the Blass and Butler techniques shows that the former needs N^2 junctions to form N beams with an N-element array, but the latter requires only $N/2 \log_2 N$. Consequently, for signal-processing uses, the Butler array is preferable. The Blass array can be made to duplicate the performance of a Butler matrix by having the number of beam inputs equal to the number of radiating elements and selecting the phases appropriately to provide orthogonality of the beams.

2.3.3 Element Spacing

For all the previously discussed array designs, element spacing is assumed to be $\lambda/2$. Closer spacing is not efficient (although arrays with close spacing, called

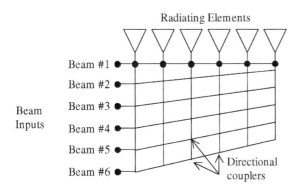

Figure 2.15 A Blass array.

super gain arrays, have been designed). Spacing that is more distant causes grating lobes (high-gain lobes) to appear in the antenna pattern at various angles. For example, for element spacing of 1λ, when the array beam is pointing on boresight (that is, all elements at the same phase), grating lobes pointing at ±90° appear because phases add up in those directions as well. For spacing of 3λ/4, grating lobes can begin to form as the amplitude of the farther out sidelobes increases. An example of array antenna patterns for element spacing and the resultant grating lobes is shown in Figure 2.16. The expression *grating lobes* comes from the diffraction gratings of physics, which are analogous but at much higher frequencies (6×10^{13} Hz vs. $\sim 10^9$ Hz).

The grating lobes are formed by a combination of the off-boresight steering angle and the element separation, as given in Equation (2.29). Consequently, the radar designer selects the two such that no grating lobes are formed during any part of the radar's beam-steering schedule. An example of array antenna patterns with λ/2 element spacing for different off-boresight steering angles and the resultant grating lobes are shown in Figure 2.17. For most surveillance radars, off-boresight steering of about 60° is provided. The accompanying element separation is 0.55λ. When element gain is greater than π (that which goes with a λ/2 aperture dimension), element separations can be correspondingly

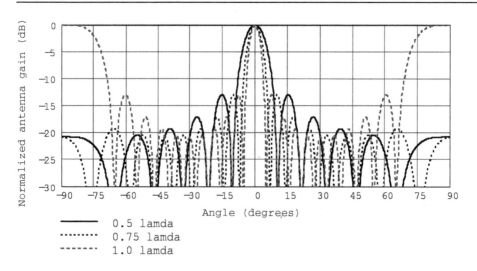

Figure 2.16 Array antenna grating lobes as a function of element spacing.

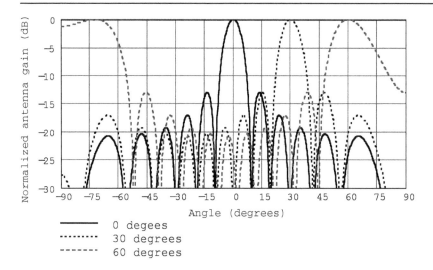

Figure 2.17 Array antenna grating lobes as a function of scan angle.

larger. Grating lobes can also be controlled by the element gain pattern. By placing nulls of the element gain pattern at the angle of the grating lobe, usually ±90°, the effect of the grating lobe can be minimized. Although a big deal is often made of grating lobes, a well designed array does not exhibit grating lobes over its designed performance (i.e., scan limits, and so on). If grating lobes were a problem, the radar's performance would certainly suffer.

$$\left| \sin(\theta_g) - \sin(\theta_0) \right| = \pm n \frac{\lambda}{d} \tag{2.29}$$

where:

θ_g = angle at which a grating lobe will appear, radians or degrees
θ_0 = beam steering angle, radians or degrees
n = integer, n = 0, 1, 2, ...
λ = wavelength, meters
d = element physical separation, meters

Array antenna gain patterns can be better understood by breaking down the array antenna gain pattern into factors, as shown in Figure 2.18. The antenna gain pattern is a function of the pattern of its elements and the array of ele-

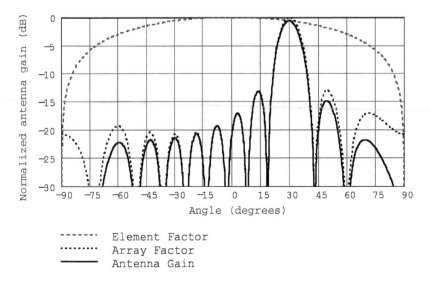

Figure 2.18   Element factor, array factor, and resultant antenna gain.

ments. The pattern of its elements is called the *element factor.* The result of the array of elements in called the *array factor.* The antenna beam can be pointed in any direction inside the element coverage. Steering a large array to thousands of beam positions per second requires the allocation of considerable computing power—no longer a problem in the modern world of computational plenty. For fixed scan patterns, the phase shifts used for the desired beam positions can be computed offline, stored in memory, and read out as needed.

2.3.4   Array Gain and Beamwidth

Several additional design approximations are used for phased arrays. Because they are made up of N elements, their boresight antenna gain is given in Equation (2.30). Often, each element is assumed to be isotropic. The antenna gain is given in Equation (2.31). Each element can have antenna gain, however. For example, crossed half-wave dipoles, or their equivalent, are often used as element antennas. Using the previously developed equation for antenna gain, the antenna gain of a half-wave dipole on a ground plane is given in Equation (2.32). Thus, the maximum antenna gain for an array of N half-wave dipoles on a ground plane is given in Equation (2.33).

$$G = N g \tag{2.30}$$

$$G = N \tag{2.31}$$

$$g_x = \frac{4\pi A}{\lambda^2} = \frac{4\pi}{\lambda^2}\left(\frac{\lambda}{2}\right)^2 = \pi \tag{2.32}$$

$$G = N \pi \tag{2.33}$$

where:
  G = boresight antenna gain for an array antennas, no units
  N = number of array elements, no units
  g = antenna gain of each element, no units
  g_x = antenna gain of a half-wave dipole on a ground plane, no units

Given appropriate element spacing, array radars obey all the general rules for antenna design that were developed for reflectors. The main difference to bear in mind is that the physical aperture is the projected area of the antenna as the beam is steered off boresight. Thus, at 60° off boresight, the aperture is down to one-half ($\cos(60°) = 0.5$) its maximum. Because the effective area of the array decreases as the beam is scanned, the antenna gain decreases, and the beamwidth increases. The effective area of the array decreases proportionally to the cosine of the beam scan angle, and thus the antenna gain decreases in the same manner [Equation (2.34)]. Thus, when scanned 60° the antenna gain of the array will be half the boresight gain. The beamwidth increases approximately inversely proportional to the cosine of the beam scan angle [Equation (2.35)]. Thus, when scanned 60°, the beamwidth of the array will be twice the boresight beamwidth. This approximation is valid only when the beam is not scanned very far off broadside ($\approx$30 ~ 40°).

$$G = N g \cos(\theta_0) \tag{2.34}$$

$$\theta_{3dB}(\theta_0) \approx \frac{\theta_{3dB}}{\cos(\theta_0)} \tag{2.35}$$

where:

$\theta_{3dB}(\theta_0)$ = array antenna half-power (–3 dB) beamwidth as a function of beam scan angle, radians or degrees

θ_{3dB} = array antenna half-power beamwidth with the beam pointing at boresight, radians or degrees

2.3.5 Array Thinning

How does one control array radar sidelobes? Although the same principles apply to arrays as applied to reflectors (amplitude weighting of the illumination function is required), the mechanisms for accomplishing it are entirely different. With reflector antennas, weighting of antenna illumination is hard to avoid, as is weighting of the signal as it scans across the target. Arrays are harder to weight. On transmit, it is simply not feasible to send differing levels of transmit power to the individual elements, changing those levels each few milliseconds as the beam scans over its surveillance and tracking program. On receive, it is wasteful to design some receivers to be less sensitive than others. However, several practical steps can be taken. On transmit, the amount of power sent to blocks of elements can be adjusted, dividing the transmit array into several subarrays and tapering the power to these. On transmit and receive, the elements can simply be physically thinned; that is, the elements at the outer edges spread out more to give the effect of an amplitude taper. As long as some randomization is done so that discontinuities (and the accompanying sidelobe spikes) are avoided, array thinning works. Control of the illumination function is excellent—much better than can be achieved with a reflector (unless an array is used to feed it), and very sophisticated weighting functions can be implemented.

Other things being equal, analysis usually recommends that the transmit array be fully filled and the receive array tapered. However, when the costs of the separate transmit and receive apertures are taken into consideration, initial costs tend to outweigh improved long-term operation. Even in a single transmit-receive aperture, however, it is feasible to fill a smaller transmit aperture fully and thinly fill the receive aperture to give good angle resolution (several receive beams following the transmitter beam around) and good sidelobe characteristics.

What constitutes optimum design also depends strongly on the mission. A surveillance mission allows a wide transmit beam and encourages a large receive aperture (surveillance cares only about adequate S/N for detection); a tracking mission wants narrow beams more than it wants S/N. (Remember, tracking improves directly as beams get narrower, but only by the square root of an improving S/N.) Moreover, of course, a combined tracker-scanner would be a complicated compromise.

2.3.6 Array Design Considerations

The preceding discussions on arrays have emphasized the radiating elements, but the techniques for driving those elements with the necessary power are also vital. Figure 2.19 shows several ways of feeding the radiating elements of an array. Each has pros and cons.

The amplifier-per-element approach has the advantages that phase shifting can be done at low power, and the losses from phase shifting made up by the subsequent amplifier. The system degrades gracefully, because random failures

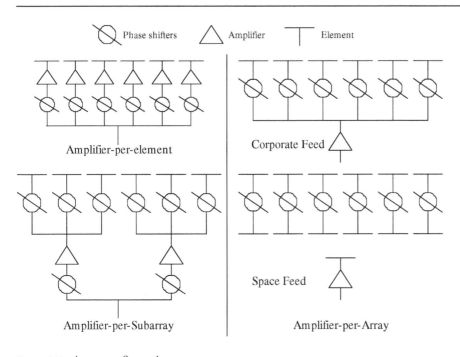

Figure 2.19 Array configurations.

of up to half the amplifiers may not appreciably affect performance. Solid-state amplifiers that will operate for 100,000 hours without failing are available [Brookner, 1985]. However, the thousands of separate amplifiers needed for even a modestly sized system increase cost and complexity. Current efforts to build a complete element (amplifier, phase shifters, and antenna) on a single chip [called a *monolithic microwave integrated circuit (MMIC)*] may result in a revolution in array design.

The amplifier-per-array approach allows the entire antenna to be driven by a single high-power tube via a network of waveguides or coaxial cables (*corporate feed*) or simply across empty space (*space feed*, or *optical feed*). The losses in the feed system and in phase shifting are not made up prior to radiation, and the phase shifting is done at high power.

In large arrays, several high-power tubes may be required. These could all be hooked in parallel behind a single feed, or each tube could be associated with a number of elements, creating several subarrays. A system made up of several subarrays is called an *array-of-subarrays.* The total number of phase shifts can now be reduced by doing coarse steering with the subarrays and fine-grained steering at the elements. If the array dimensions are large compared to the pulse widths to be radiated, the radiated pulses will be smeared when the antenna is pointed off boresight. To reduce smearing, the subarrays can be steered with time delay networks rather than phase shifters.

Reflectors and arrays with various steering features can be mixed to create the antenna design required for a particular mission. A reflector might be fed with an array of horns to form multiple beams in a fairly narrow solid angle, which is then scanned around the sky mechanically. In airborne attack radars, which must be carried on fighter aircraft where space and weight are at a premium, the antenna may be scanned mechanically in one dimension (usually azimuth) and phase steered in the other (elevation). Such an antenna may be physically flat, for ruggedness and to save space, the antenna's azimuth beam being formed by fixed phasors and its elevation beam by phase shifters.

2.4 EXERCISES

1. You are asked to design an antenna for an infrared radar (wavelength about 20 microns [micrometers, μm]) with a gain $G = 10^6$ (60 dB). Assume the an-

tenna efficiency ρ = 100%. Assuming a square antenna shape, what are the dimensions of the aperture? What size would it be in the microwave region, say at S-band (f_c = 3000 MHz)?

2. A reflector antenna has an area A = 100 m². How many maximum gain elements would there be in a fully filled UHF ($\lambda \approx 1$ m) array of equivalent capability? What would be its area?

3. Assume cosine weighting of a linear aperture of length $-\pi/2$ to $+\pi/2$. Show that $G(\theta)$ is of the form

$$\left[\frac{\sin(1-\theta)\frac{\pi}{2}}{(1-\theta)\frac{\pi}{2}} + \frac{\sin(1+\theta)\frac{\pi}{2}}{(1+\theta)\frac{\pi}{2}} \right]$$

Calculate sidelobes at θ = 4°, 6°, 8°; square them; and turn them into deci-

bels to get –23, –30, and –36 dB. (Hint: Use the identity $\cos x = \dfrac{e^{jx} + e^{-jx}}{2}$

to get the expression into integrable form.)

4. Calculate the mainbeam gain and the azimuth and elevation beamwidths for a radar antenna with the following characteristics: Uniform illumination function, 5 m wide, 2 m high, antenna efficiency ρ = 0.5, and a frequency f_c = 3 GHz.

5. Calculate the normalized gain of an antenna at an angle of 5° for an antenna with a uniform current distribution, length D = 3 m, and a frequency f_c = 5 GHz. If the mainbeam gain G = 35 dB, what is the gain of the antenna at an angle of 5°?

6. A phase-steered array has the following characteristics: frequency f_c = 6 GHz, element spacing $d = \lambda/2$, and phase shift between elements β = 15°. To what angle is the beam steered?

7. A frequency scan array has a transmit frequency f_c = 2 GHz, a frequency excursion Δf = 100 MHz, element spacing $d = \lambda/2$, and electrical distance between elements L = 5λ. What is the beam steering limit? What is the wrap-up factor, L/d, needed to provide a beam steering limit $\theta_1 = \pm 60°$?

8. An array antenna uses 3-bit phase shifters and has a beamwidth θ_{3dB} = 2°. How precisely can the beam be steered?

9. Calculate the beamwidth of an electronically scanned array with a beam-width $\theta_{3dB} = 2.5°$ when pointed at $0°$ when it is scanned to $25°$, $30°$, and $45°$.

10. Calculate the mainbeam antenna gain for an electronically scanned array with the following characteristics: number of elements $N = 1000$ and element gain $g = 1.5$ dB when it is scanned to $25°$, $30°$, and $45°$.

2.5 REFERENCES

Allen, J. L., 1963, *The Theory of Array Antennas,* Lexington, MA: Lincoln Laboratory Tech. Report #323.

Blass, J., 1960, "Multidirectional Antenna-A New Approach to Stacked Beams," 1960 IRE International Convention Record, Pt. 1, pp. 48–50.

Brookner, E., 1985, "Phased Array Radars, *Scientific American,*" Vol. 252, no. 2, Feb. 1985, p. 100.

Gardner, M., 1981, "Mathematical Games," *Scientific American,* Vol. 245, no. 2, Aug. 1981, pp. 16–26.

Hansen, R. C., 1966, Microwave Scanning Antennas, Vol. 2., New York: Academic Press.

Moody, H. J., 1964, "The Systematic Design of the Butler Matrix," *IEEE Transactions,* Vol. AP-1 I, Nov., pp. 786–788.

Shelton, J. W., 1968, "Fast Fourier Transforms and Butler Matrix," *Proc. IEEE,* Vol. 56, Mar., p. 350.

Skolnik, Merrill I., 2001, *Introduction To Radar Systems,* 3rd ed., New York: McGraw-Hill. Skolnik has a great deal of additional information on both reflector and array antennas in Chapter 9, "The Radar Antenna."

Stark, L., 1974, "Microwave Theory of Phased Array Antennas-A Review," *Proc. IEEE,* Vol. 62, Dec., pp. 1661–1701.

Stark, L., R. W. Burns, and W. P. Clark, 1970, "Phase Shifters for Arrays," in M. I. Skolnik (Ed.), *Radar Handbook,* New York: McGraw-Hill.

Stimson, George W., 1998, *Introduction To Airborne Radar,* 2nd ed., Raleigh, NC: SciTech Publishing. Stimson has a great deal of additional information on array antennas in Chapters 37 and 38.

Swerling, P., 1954, "Probability of Detection of Fluctuating Targets," Rand Corporation Research Memo RM-1217. (Reprinted *IRE Transactions,* IT-6, Apr. 1960.)

Temes, C. L., 1962, "Sidelobe Suppression in a Range-Channel Pulse Compression Radar," *IRE Transactions,* Vol. MIL-6, Apr., pp. 162–167.

Temme, D. H., 1972, "Diode and Ferrite Phase Shifter Technology," in A. A. Olner and G. H. Knittel (Eds.), *Phased Array Antennas,* Dedham, MA: Artech House.

3

Detection and Tracking

Assume that radars having some of the parameters described in the preceding chapters can be built and operated. A vitally important question is how the operator decides when he sees a target. He has found out quickly from experience that he can turn up the gain on his radar display and completely fill his field of view with false targets (noise), or he can turn down the gain and eliminate all targets. How does he decide where to set the threshold to optimize his ability to detect? Likewise, how do we set the threshold for the automatic detection systems used by the majority of modern radar systems?

3.1 THE PROBLEM OF DETECTION

An experienced operator becomes skillful at setting the gain on a radar display. Fortunately, there is also a quantitative way to evaluate what the settings should be so that targets can be detected automatically and the operator informed by the simple expedient of a light turning on or a computer trigger circuit activating. The problem of detection is therefore reduced to finding an appropriate threshold, one above which noise samples seldom rise and below which signal pulses seldom fall.

3.2 NOISE DISTRIBUTIONS

To get at detection quantitatively, one must assume some characteristics for the noise that appears at the output of a radar receiver. At most radar frequencies, the principal contributor has been found to be receiver thermal noise. Prior to the envelope detector (rectification), which is energy detection in the electronic circuits, receiver thermal noise has a Gaussian (or normal) distribution with zero mean. (Energy detection is not analogous to target detection, because the detector circuits in a radar receiver also "detect" noise. Detection circuits are best thought of as "thresholding.") After rectification, receiver noise has a one-sided probability distribution (negative values are not possible), fluctuating about a mean that is the root of the mean-squared value of the unrectified fluctuations. A distribution that is easy to handle and satisfies these constraints (and most of the other required ones) is the exponential distribution. In slightly altered form, it is also called the Rayleigh distribution as well as being a member of the family of gamma distributions [Freund, 1962, pp. 127–128]. The exponential distribution is given in Equation (3.1) and shown in Figure 3.1.

$$f(x) = \frac{1}{\theta} e^{-x/\theta} \qquad x > 0$$

$$= 0 \qquad\qquad \text{elsewhere} \tag{3.1}$$

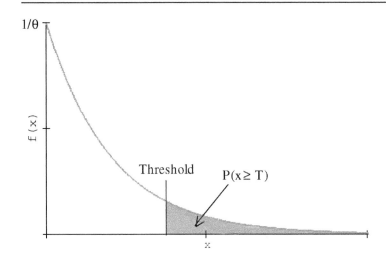

Figure 3.1 Exponential probability distribution.

where:

$f(x)$ = probability of x occurring for the exponential probability distribution

θ = mean and standard deviation of the exponential probability distribution

θ^2 = variance of the exponential probability distribution

For the radar receiver noise, θ is the RMS noise power, the N we discussed in Chapter 1, and x is the peak noise power in a sample of noise. The fact that the exponent of x is one (1) in the exponential distribution makes it very convenient for calculations. However, it places constraints on us. The x must represent power, not voltage. Voltage would have to be represented by x^2 (making calculations more difficult). Our distribution is applicable only to square-law detectors for which the receiver responds linearly to signal power [Barton, 1979, p. 12]. Square-law detectors are often used, however, in radar receivers.

There are a few mathematical relationships that we take advantage of for radar detection. First, if we integrate a probability distribution over its range of values we get the value one (1). Because the rectified noise is greater than zero, this mathematical relationship is given in Equation (3.2). Second, the probability of exceeding a specific value is given in Equation (3.3). This probability is also shown in Figure 3.1. We will use this mathematical relationship for radar detection.

$$\int_{-\infty}^{\infty} f(x)\,dx = 1 \qquad\qquad \int_{0}^{\infty} f(x)\,dx = 1 \tag{3.2}$$

$$P(x \geq T) = \int_{T}^{\infty} f(x)\,dx = 1 - \int_{0}^{T} f(x)\,dx \tag{3.3}$$

where:

$f(x)$ = probability distribution

$P(x \geq T)$ = probability of x exceeding the value T

It is easy to calculate the probability that noise will exceed a given threshold using the exponential distribution. We can use Equations (3.3) and (3.1) to calculate the probability of noise exceeding a threshold T, as given in Equation (3.4). For example, the probability of noise exceeding a threshold ten times the mean of the exponential distribution is given in Equation (3.5). The probability that the noise will cross a given threshold is called the *probability of false alarm* (P_{fa}) as in Equation (3.6). From a statistical perspective, the probability of false alarm is the *conditional* probability that given no signal is present; the noise alone will exceed the detection threshold. Equation (3.6) can be solved in terms of the probability of false alarm, Equation (3.7). For example, a probability of false alarm of 10^{-6} requires a threshold-to-noise ratio of 13.8 (11.4 dB) [Equation (3.8)].

$$P(x \geq T) = \int_{T}^{\infty} \frac{1}{N} e^{-x/N}\,dx = -e^{-x/N}\Big|_{T}^{\infty}$$

$$= 0 - (-e^{-T/N}) = e^{-T/N} \tag{3.4}$$

$$P(x \geq 10N) = e^{-10N/N} \cong 4.54 \times 10^{-5} \tag{3.5}$$

$$P_{fa} = e^{-T/N} \tag{3.6}$$

$$\frac{T}{N} = -\ln(P_{fa}) \tag{3.7}$$

$$\frac{T}{N} = -\ln(10^{-6}) = 13.8 \qquad 10 \log(13.8) = 11.4 \text{ dB} \tag{3.8}$$

where:
   $P(x \geq T)$ = probability of noise exceeding a threshold, no units
      T = threshold value, watts
      N = RMS noise power, watts
      P_{fa} = probability of false alarm
   T/N = threshold-to-noise ratio, no units

Often, probability of false alarm does not provide a tangible indication of false alarms. The average time between false alarms can provide this tangible indication. The average time between false alarms is a function of the probability of false alarm and the receiver bandwidth [Equation (3.9)]. When we combine Equations (3.9) and (3.6), we obtain a relationship between the average time between false alarms and the threshold-to-noise ratio [Equation (3.10)]. A plot of the average time between false alarms as a function of the threshold-to-noise ratio and receiver bandwidth is shown in Figure 3.2.

$$T_{fa} = \frac{1}{P_{fa} B} \tag{3.9}$$

$$T_{fa} = \frac{1}{B} e^{T/N} \tag{3.10}$$

where:
   T_{fa} = average time between false alarms, seconds
      B = receiver bandwidth, hertz

3.3 SIGNAL-TO-NOISE RATIO

So far, we have said nothing about the signal. The signal almost invariably has different statistics from the receiver noise. We can start by ascribing to it a constant amplitude. Its probability distribution is an impulse, as shown in

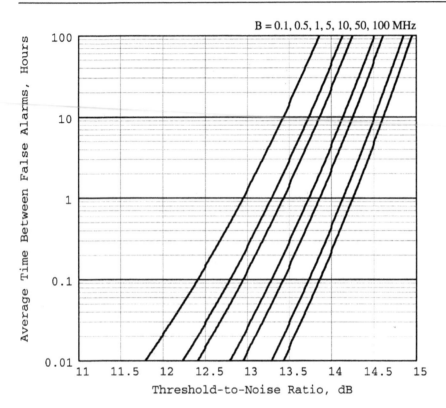

Figure 3.2 T_{fa} as a function of T/N and receiver bandwidth.

Figure 3.3. In a noiseless environment, it would appear with value S (target signal power from Chapter 1) at a calculable location, and detection thresholding would become elementary. Unfortunately, the signal has added to it the noise admitted into and generated by the receiver. The resulting probability distribution is the convolution of the two, which, because the signal distribution is just an impulse, becomes the noise distribution displaced to the right by the signal amplitude. However, when the signal is present, the exponential distribution no longer applies, because values below S, as well as above it, are now permitted. By the previous description of predetection noise, we know that we have ascribed Gaussian characteristics to it. Now, however, instead of fluctuating about a zero mean, it will have a mean at S, the RMS signal power. If $S \gg N$, which it should be for the radar to operate effectively, we are far enough away from zero

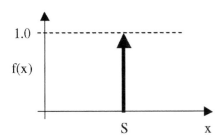

Figure 3.3 Impulse probability distribution for the signal.

that the inaccuracies associated with ignoring the negative values will be negligible. It seems reasonable, therefore, to use the Gaussian distribution to describe signal-plus-noise. North (1943) has shown that the voltage of the signal-plus-noise does follow the Gaussian probability distribution, which is contained in many books of mathematical tables.

3.4 DESIGNING DETECTION THRESHOLDS

Where previously the idea was to set a threshold for an arbitrary probability that noise alone would exceed it, now the threshold may be placed to be sure that the signal-plus-noise exceeds it for an arbitrary probability. Presume that a signal processor designer is thinking in terms of a $\geq 90\%$ probability of detection (P_d) for a single pulse. This simply means that his threshold should be set >1.28 standard deviations below the mean of a Gaussian signal-plus-noise probability distribution. For a 99.9% P_d, three standard deviations below the mean are necessary. The probability of detection is shown in Figure 3.4. From a statistical perspective, the probability of detection is the *conditional* probability that given a signal is present, the signal-plus-noise exceeds the detection threshold.

A choice of detection threshold based on P_{fa} has been discussed. So has one based on P_d. The actual threshold chosen must be based on a trade-off of the two with the radar capability. Given an existing radar, the achievable signal-to-noise ratio, S/N, is determined, and thresholds are then calculated based on the mission. Or if a mission is defined and a P_{fa} and P_d assigned (and a threshold thereby implicitly defined), the radar must then be designed to achieve the

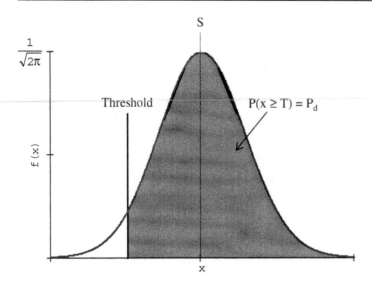

Figure 3.4   Gaussian probability distribution and probability of detection.

S/N required. With the simple theory previously developed, both operations can be conducted, as shown in Figure 3.5. The P_d is found by integrating the Gaussian probability distribution based on a *S/N* threshold that obtains a desired P_{fa}. Unfortunately, the integral of the Gaussian probability distribution does not result in a closed-form solution. An accurate approximation to this integral was presented by North, 1963, Equation (3.11), based on the complementary error function, Equation (3.12).

$$P_d \approx 0.5 \text{ erfc}\left(\sqrt{-\ln(P_{fa})} - \sqrt{\frac{S}{N} + 0.5} \right) \tag{3.11}$$

$$\text{erfc}(T) = 1 - \frac{2}{\sqrt{\pi}} \int_0^T e^{-x^2} dx \tag{3.12}$$

Detailed curves of the relationship between P_d, P_{fa}, and *S/N* have been worked out by many [Barton, 1988, p. 62; Marcum, 1960; Skolnik, 2001, p. 44;

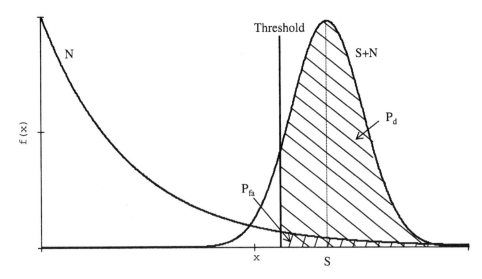

Figure 3.5 Probability of detection and probability of false alarm.

DiFranco and Rubin, 1980]. We have used Equation (3.11) to generate such curves. The result is shown in Figure 3.6 and tabulated in Table 3.1.

If the return signal cannot be represented by an impulse probability distribution (it may have modulation added to it by either a fluctuating target RCS or the intervening medium or both), then the probability distribution of $S + N$ is more dispersed, and its standard deviation is substantially greater. From elementary statistics, we know that, for independent random variables, the variance of the sum of these distributions will be the sum of the variances. Accordingly, considerably higher S/Ns are required for high P_d, but the fluctuating target has better P_d at low S/Ns. A great deal of work has been done to understand radar detection theory for fluctuating target signals. The most widely used is that performed by Dr. Peter Swerling [Skolnik, 2001, pp. 66–70; DiFranco and Rubin, 1980, Chapter 11], which is well documented in the above radar detection references. A great deal of work has been done in further refining detection theory by classifying targets, postulating the optimum radar scan, and signal-processing techniques to guarantee some combination of P_{fa} and P_d. When a target's RCS fluctuates with *known* characteristics, the radar can be designed to detect only the higher RCS, ignoring the low RCS returns.

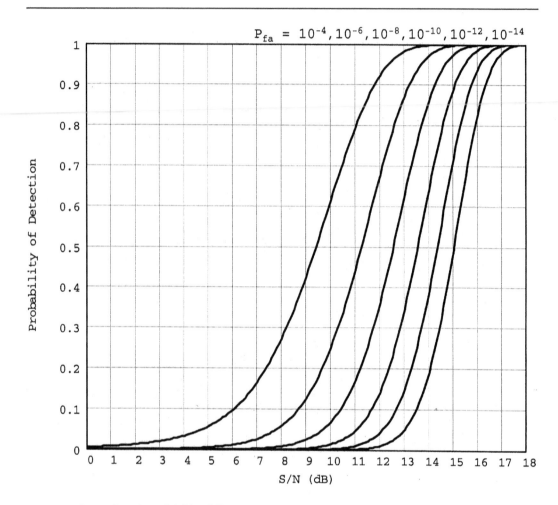

Figure 3.6 P_d as a function of S/N and P_{fa}.

However, if the RCS is probabilistic, the radar must be designed to detect it most of the time.

The thresholding exercises previously discussed have an exact analogy in mathematical statistics, where they are called *hypothesis tests*. There, the P_{fa} is called the *probability of making a "β error,"* that is, accepting a hypothesis falsely, and $1 - P_d$ is the probability of making an "α error," that is, rejecting a hypothesis erroneously [Freund, 1962, pp. 239–240].

Table 3.1   S/N (dB) Required for Detection as a Function of P_d and P_{fa}

					P_{fa}						
P_d	10^{-4}	10^{-5}	10^{-6}	10^{-7}	10^{-8}	10^{-9}	10^{-10}	10^{-11}	10^{-12}	10^{-13}	10^{-14}
0.05	4.8	6.5	7.8	8.8	9.7	10.4	11.0	11.6	12.1	12.6	13.0
0.10	6.1	7.5	8.7	9.6	10.4	11.1	11.7	12.2	12.7	13.1	13.5
0.15	6.8	8.2	9.2	10.1	10.9	11.5	12.0	12.6	13.0	13.4	13.8
0.20	7.4	8.7	9.7	10.5	11.2	11.8	12.3	12.8	13.3	13.7	14.0
0.25	7.8	9.0	10.0	10.8	11.5	12.1	12.6	13.1	13.5	13.9	14.2
0.30	8.2	9.4	10.3	11.1	11.7	12.3	12.8	13.3	13.7	14.1	14.4
0.35	8.5	9.7	10.6	11.3	11.9	12.5	13.0	13.5	13.9	14.2	14.6
0.40	8.8	9.9	10.8	11.5	12.2	12.7	13.2	13.6	14.0	14.4	14.7
0.45	9.1	10.2	11.0	11.7	12.3	12.9	13.4	13.8	14.2	14.5	14.9
0.50	9.4	10.4	11.2	11.9	12.5	13.1	13.5	13.9	14.3	14.7	15.0
0.55	9.7	10.7	11.5	12.1	12.7	13.2	13.7	14.1	14.5	14.8	15.2
0.60	9.9	10.9	11.7	12.3	12.9	13.4	13.9	14.3	14.0	15.0	15.3
0.65	10.2	11.1	11.9	12.5	13.1	13.6	14.0	14.4	14.8	15.1	15.4
0.70	10.5	11.4	12.1	12.7	13.3	13.8	14.2	14.6	14.9	15.3	15.6
0.75	10.7	11.6	12.3	12.9	13.5	13.9	14.4	14.8	15.1	15.4	15.7
0.80	11.0	11.9	12.6	13.2	13.7	14.1	14.6	14.9	15.3	15.6	15.9
0.85	11.4	12.2	12.9	13.4	13.9	14.4	14.8	15.2	15.5	15.8	16.1
0.90	11.8	12.5	13.2	13.8	14.2	14.7	15.1	15.4	15.7	16.0	16.3
0.95	12.3	13.1	13.7	14.2	14.7	15.1	15.4	15.8	16.1	16.4	16.7
0.99	13.3	14.0	14.5	15.0	15.4	15.8	16.1	16.4	16.7	17.0	17.3

3.5   SOME DETECTION TECHNIQUES

3.5.1   Integration

The primary method of enhancing the detection performance of a radar is to use multiple pulses to "dig" the signal out of the noise. In this operation, multiple pulses can be sent to all resolution bins and the energy added in all. The resultant summing is called *integration.* In implementing integration of multiple received pulses, both signals and noise are stored with their magnitudes and phases. This may be done in a computer or a delay line. As ensuing returns arrive, they are added. Accumulated stored returns that exceed the detection threshold are declared as detections.

Is the technique of integration any different from sending the same power in a short pulse? Theoretically, it is not, but there are losses associated with integration in mismatches due to elapsed time. In general, however, multiple-pulse integration is more economical than generating a single pulse of the same total energy.

There are two main types of integration: *coherent* and *noncoherent.* Coherent integration is also called *predetection integration,* because it takes place before the envelope detector using both the magnitude and phase terms of the received pulses and receiver noise samples *(S + N).* Noncoherent integration is also called *postdetection integration,* because is takes place after the envelope detector using only the magnitude of the *S + N.* For coherent integration, a coherent (all the pulses are in phase with each other) waveform is transmitted, and thus the received pulses are coherent. Thus, the amplitude of each pulse simply adds (in voltage or increases quadraticaly in power) in phase with the other pulses, as given in Equation (3.13). The receiver thermal noise samples are random signals and thus add as complex (magnitude and phase) random numbers, as given in Equation (3.14). Combining these two equations shows how coherent integration provides a linear increase in the *S/N,* as given in Equation (3.15).

$$S_n = n_p^2 \, S_1 \tag{3.13}$$

$$N_n = n_p \, N_1 \tag{3.14}$$

$$\left(\frac{S}{N}\right)_n = \frac{n_p^2 \, S_1}{n_p \, N_1} = n_p \left(\frac{S}{N}\right)_1 \tag{3.15}$$

where:

S_n = received target signal power after integration of n_p coherent pulses, watts

n_p = number of pulses integrated

S_1 = received target signal power from one pulse, watts

N_n = receiver thermal noise power after integration of n_p noise samples, watts

N_1 = one sample of receiver thermal noise power, watts

$(S/N)_n$ = signal-to-noise ratio after integration of n_p coherent pulses, no units

$(S/N)_1$ = single pulse signal-to-noise ratio, no units

It is difficult and expensive to build a coherent radar system (transmitter, receiver, and signal processor), especially a high-power transmitter. Thus, the majority of radar systems use a noncoherent waveform. With a noncoherent waveform, the received pulses are also not necessarily in phase with each other. They are correlated (from the same source and reflected from the same target) signals, not purely random signals. Thus, they add as complex (magnitude and phase) correlated numbers. The improvement in received target signal power due to noncoherent integration is a complicated function of P_d, P_{fa}, and target RCS fluctuation characteristics [Skolnik, 2001, p. 65].

To account for the improvement in S/N due to integration of multiple pulses, radar engineers often use an integration gain term, as given in Equation (3.16). For coherent integration, the integration gain is given in Equation (3.17). For the noncoherent integration, the range of integration gain is given in Equation (3.18), with a commonly used rule of thumb given in Equation (3.19). Specific values for noncoherent integration gain can be found in the above radar detection references. Marcum (1960) states the optimum noncoherent integration gain is approximately the number of pulses integrated raised to the 0.76 power, as given in Equation (3.20). These integration gain terms are shown in Figure 3.7.

$$\left(\frac{S}{N}\right)_n = G_I \left(\frac{S}{N}\right)_1 \qquad (3.16)$$

$$G_I = n_p \quad \text{Coherent integration} \qquad (3.17)$$

$$\sqrt{n_p} < G_I < n_p \quad \text{Noncoherent integration} \qquad (3.18)$$

$$G_I \approx \frac{n_p + \sqrt{n_p}}{2} \quad \text{Noncoherent integration } \textit{rule of thumb} \qquad (3.19)$$

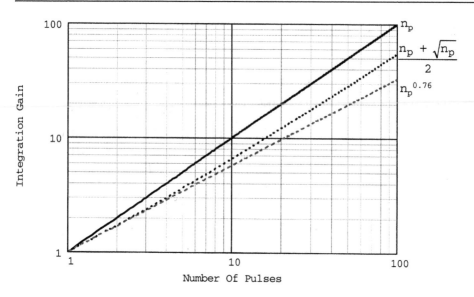

Figure 3.7 Integration gain—coherent and noncoherent.

$$G_I \approx n_p^{0.76} \quad \text{Noncoherent integration } \textit{Marcum} \tag{3.20}$$

where:

G_I = integration gain, no units

The multiple pulse signal-to-noise ratio is the product of the single pulse signal-to-noise ratio [Equation (1.14)] and the integration gain, as given in Equation (3.21).

$$\left(\frac{S}{N}\right)_n = \frac{P\,G^2\,\lambda^2\,\sigma\,G_I}{(4\pi)^3\,R^4\,(F_n-1)\,k\,T_0\,B\,L_s} \tag{3.21}$$

The number of pulses integrated is a function of the integration time and the *PRF*, as given in Equation (3.22). The integration time is often a fixed value based on radar signal processing conditions. This is the case for pulse Doppler and tracking radar systems. The integration time can also be the target illumination time, as given in Equation (2.21). This is the case for scanning radar systems. The target illumination time is the time that the target is within the radar antenna's 3-dB beamwidth. The maximum number of pulses available for integration for scanning radar is computed using the target illumination time. Some scanning radars integrate all the pulses in one target illumination time, whereas others integrate only a portion of the pulses.

$$n_p = T_I \ PRF \tag{3.22}$$

$$T_{ill} = \frac{\theta_{3dB}}{\theta_{dot}} \tag{3.23}$$

where:
T_I = integration time, seconds
PRF = pulse repetition frequency, hertz
T_{ill} = target illumination time, seconds
θ_{3dB} = antenna 3-dB beamwidth, degrees or radians
θ_{dot} = antenna scan rate, degrees/second or radians/second

The multiple pulse radar range equation can be solved for the range at which the radar will detect the presence of a target with a given S/N. The radar detection range is given in Equation (3.24).

$$R_d = \sqrt[4]{\frac{P \ G^2 \ \lambda^2 \sigma \ G_I}{(4\pi)^3 \ SNR_d \ (F_n - 1) k \ T_0 \ B \ L_s}} \tag{3.24}$$

where:
R_d = radar detection range, meters
SNR_d = signal-to-noise ratio required for detection (detection threshold), no units

3.5.2 Cumulative Probability of Detection

As the target approaches the radar, the S/N increases, and thus the individual probability of detection increases. A cumulative probability of detection is the combination of multiple individual probabilities of detection (from each scan or tracking measurement). The cumulative probability of detection is given in Equation (3.25). The cumulative probability of false alarm can also be computed using this equation by simply replacing the probability of detection with the probability of false alarm. It is common to see detection ranges defined for a certain cumulative probability of detection, especially for airborne radar systems. Care must be taken in interpreting such a detection range, because the equation for cumulative probability of detection is based on all individual probabilities of detection being associated (i.e., correlated to use the statistical term) with the same target. This correlation is easy to ensure in theory, but very hard to ensure in practice.

$$P_{dc} = 1 - \prod_{i=1}^{n} (1 - P_{di}) \tag{3.25}$$

where:

P_{dc} = cumulative probability of detection
n = number of individual detections (scans or measurements) combined
P_{di} = probability of detection for the ith individual detection

3.5.3 Sequential Detection

Another method of enhancing detection performance, if system time lines allow it, is to use more than one pulse and/or detection decision to determine whether a target is present. This can be done in many ways, but most such tactics are really probability games. One may set a threshold (T_1) at relatively low S/N where P_{fa} is high. Range and angle resolution bins in which threshold crossing occurs are noted, and an additional pulse (sometimes called a *verification pulse*) is sent. If T_1 is crossed again in the same locations, detection is declared. Note what has happened to the detection statistics because of sending the second pulse. The probability of noise alone exceeding the detection threshold twice is markedly lower than just once, as given in Equation (3.26).

$$P_{fa}(2) = P_{fa}(1)^2$$

$$P_{fa}(n) = P_{fa}(1)^n \tag{3.26}$$

where:

$P_{fa}(2)$ = probability of noise alone exceeding the detection threshold twice
$P_{fa}(1)$ = probability of noise alone exceeding the detection threshold once (P_{fa})
$P_{fa}(n)$ = probability of noise alone exceeding the detection threshold n times

The utility of this approach is twofold. It permits an improvement of radar performance (at a small sacrifice in time and radar complexity) without adding power aperture, and it makes for more efficient use of existing capability, because no verification pulse is sent when T_1 is not crossed. We must take care, however, in interpreting what this two-pulse thresholding scheme means. The original threshold was set to make the P_d as high as possible at the expense of letting the P_{fa} also be relatively high. The level of the threshold relegates a certain percentage of targets to remaining undetected on average. The second pulse does not allow us to increase P_d (a posterior that is 1.0, given that the signal-plus-noise that has already crossed the threshold, is not noise alone). What the second pulse does allow us to do is reduce the P_{fa} by the square of its single pulse value; for example, if the chance that a noise sample crossed the threshold once were 1 in 1000, the chance that it would cross twice would be 1 in 1 million.

3.5.4 *M*-out-of-*N* Detection

One multiple-pulse detection scheme uses the binomial probability theorem. A threshold is established, and an event is based on whether that threshold is crossed. The probability of crossing the detection threshold (a detection) N times out of M attempts is given in Equation (3.27).

$$P(M,N,p) = \frac{N!}{M!\,(N-M)!}\, p^M (1-p)^{(N-M)} \tag{3.27}$$

where:
  $P(M,N,p)$ = probability of M detection threshold crossing out of N attempts
    M = number of times the detection threshold is crossed
    N = number of attempts
    p = probability of crossing the detection threshold for each attempt

For simplicity, let us consider a two-attempt binomial detection. If the threshold is set to achieve a $P_d = 0.5$ (that is, $p = 0.5$), the possible outcomes are two threshold crossings with probability 0.25, one threshold crossing with probability 0.5, and zero threshold crossings with probability 0.25. The calculations for this example are shown in Equation (3.28). The probability of at least one threshold crossing is $0.25 + 0.5 = 0.75$. Therefore, we have enhanced P_d by 50% using a two-attempt binomial detection.

$$P(2,2,0.5) = \frac{2!}{2!(2-2)!}(0.5)^2(1-0.5)^{(2-2)} = 0.25$$

$$P(1,2,0.5) = \frac{2!}{1!(2-1)!}(0.5)^1(1-0.5)^{(2-1)} = 0.5$$

$$P(0,2,0.5) = \frac{2!}{0!(2-0)!}(0.5)^0(1-0.5)^{(2-0)} = 0.25$$

$$(3.28)$$

What about the probability of false alarm? Assume that the detection threshold is set to achieve a $P_d = 0.5$ with a $P_{fa} = 10^{-4}$. Again, there are three outcomes of the two-attempt set: two threshold crossings with probability 10^{-8}, one threshold crossing with probability 1.9998×10^{-4}, and zero threshold crossings with probability 0.9998. The probability of at least one threshold crossing is $10^{-8} + 1.9998 \times 10^{-4} = 1.9999 \times 10^{-4}$, which means that the P_{fa} has approximately doubled.

Referring to Table 3.1, the detection threshold to achieve a $P_d = 0.5$ with a $P_{fa} = 10^{-4}$ is 9.4 dB. The detection threshold to achieve a $P_d = 0.75$ with a $P_{fa} = 2 \times 10^{-4}$ is 10.4 dB. If a probability of at least one threshold crossing is desired to be 0.5, then the probability of detection of each attempt would be approximately 0.3. The detection threshold to achieve a $P_d = 0.3$ with a $P_{fa} = 2 \times 10^{-4}$ is 7.8 dB. This reduction in the required detection threshold, while maintaining a desired P_d and P_{fa}, is one of the values of binomial detection.

Imagine a five-attempt binomial detection with a three-out-of-five threshold-crossing criterion. We retain single attempt $P_d = 0.5$, $P_{fa} = 10^{-4}$, and S/N = 9.4 dB. The probability of three or more threshold crossings when single-attempt $P_d = 0.5$ is 0.5. For this criterion, the probability of detection remains unchanged. For single-attempt $P_{fa} = 10^{-4}$, the probability of three or more threshold crossings is 9.9985×10^{-12}. (The steps to achieve these numbers are shown in Exercise 6.) For a three-out-of-five criterion, we have succeeded in lowering the probability of false alarm by seven orders of magnitude while keeping the probability of detection unchanged. This reduction in the probability of false alarm, while maintaining a desired P_d and detection threshold, is one of the values of binomial detection. Once again referring to Table 3.1, the detection threshold to achieve $P_d = 0.5$ with a $P_{fa} = 10^{-11}$ is 13.9 dB.

M-out-of-*N* detection criteria are used in many surveillance radars, because they are easy to implement in a computer and allow reductions in the probability of false alarm at constant, or even improving, probabilities of detection with modest detection thresholds. The P_d and P_{fa} can be manipulated rather easily as well. To raise P_d, minimize the number of crossings required for detection; to lower P_{fa}, do the opposite.

There are other, statistically sophisticated, multiple-threshold detection concepts, some of which are done in a computer rather than the radar. Despite initial enthusiasms, the additional enhancement of performance over the simple methods already described is probably less than 3 dB. (The impact on detection range of such a change is illustrated in Exercise 8.)

3.6 TRACKING

Once a target has been detected, it can be followed in range, angle, range rate (Doppler), and time. It can either be actively followed by the radar in a special tracking mode, or its whereabouts can simply be noted, by either an operator or computer, each time it is detected on a radar scan. The latter technique is called *track-while-scan* [Barton, 1979, pp. 82–86; Skolnik, 2001, pp. 212, 252–254; Stimson, 1998, pp. 388–390] and is the method used by most FAA radars, which provide synthetic presentations to the scopes of the air traffic controllers. A phased array radar, with its inertialess, electronically steered beams, can maintain a routine scan while actively tracking many targets in a mode of operation known as *track-and-scan.*

The track-while-scan mode is an operation involving correlations and extensive calculations but virtually no radar operations. we will not discuss it further.

Because their ability to track targets at varying angles is physically limited, most single-beam reflector radars have only a few tracking channels (one to four). A phased array may have very many (several hundred). The operator of a flight control radar (track-while-scan) may carry tens of tracks in his head. Radar tracking capabilities vary widely and depend on the mission to be accomplished and the radar design.

Active tracking requires several important radar operations. Radars can track in angle, range, and Doppler. They often track in all three, and all three deserve discussion.

3.6.1 Angle Tracking

Methods for following a target in angle have evolved from the primitive approach of having the tracker seek a peak return, through conical scan techniques, to monopulse.

3.6.1.1 *Conical scan techniques.* In conical scan, the antenna feedhorn is made to rotate around a boresight axis at a prescribed rate, which is slower than the *PRF* but faster than the antenna's angular rate [Skolnik, 2001, pp. 225–229; Stimson, 1998, pp. 9 and 102]. The rotation causes the antenna beam to describe a small-angle cone in space. Figure 3.8 is a cross section of the far-field beams. The returns from the pulses during a scan are compared in amplitude, and the antenna is moved toward the position where they would balance. In the meantime, returns from a new scan are coming in.

Conical scan radar systems are still used because of their simplicity, but they have many limitations:

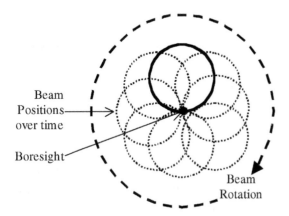

Beam
Positions
over time

Boresight

Beam
Rotation

Figure 3.8 Far-field conical scan pattern.

1. The target *RCS* can change during a scan interval, resulting in angle errors due to *RCS* fluctuations,

2. Amplitude comparisons are not particularly accurate.

3. The conical scan is very susceptible to electronic countermeasures, because the scan rate is readily detectable in the radar signal and can be contaminated to confound the tracker's servomechanisms.

3.6.1.2 *Monopulse tracking*. Angle tracking in modern radars is usually done with a scheme known as *monopulse tracking* [Barton, 1974; Rhodes, 1959; Skolnik, 2001; Stimson, 1998]. All the limitations of conical scan tracking are eliminated in monopulse. The term is used because the information obtained from a single pulse tells the radar where the object is in angle, how far the beam is from pointing at the target exactly, and provides an error signal that drives the radar beam toward the target. Good monopulse trackers can easily get accuracies of one-tenth of the radar beamwidth. In fact, they can come close to achieving the theoretical angular accuracy capability of a radar.

Monopulse tracking is essentially simple. The feedhorns of a monopulse tracker are slightly displaced so that each receives the signal from a slightly different angle. The received beams can be added to form a sum beam (which is used for gross pointing) and subtracted to form a difference beam. Wherever the crossover of the two beams is selected, the difference beam will produce a zero response when the target is on the antenna boresight. If the difference beam is not zero, the deviation is used to generate a voltage that drives the beam toward zero or toward the so-called monopulse null. The technique is portrayed in two dimensions in Figure 3.9. Either amplitude or phase information may be used. The difference beam curve is linear near the origin. The amplitude of the correcting voltage developed is proportional to the distance from the null, and its sign indicates the direction to the null. A servomechanism drives the pedestal, rotating the antenna toward the null. To extend the case to two angles, a third horn is added, and separate comparisons are made. There are also four-horn and even five-horn monopulse systems.

In a phased array radar, the identical technique can be employed by using phase changes in the beam-steering circuitry instead of a pedestal servo drive. Often, however, phased array beams are not moved toward the monopulse null.

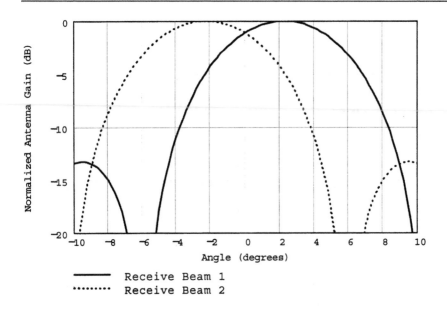

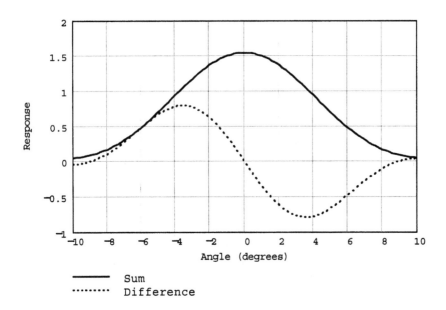

Figure 3.9 Monopulse tracking.

Instead, the position of the target is extrapolated from the size of the signal generated in the difference beam output.

3.6.2 Range Tracking

Range tracking implies following an object in range as a function of time. A "tracking gate" is put around an object designated for range track. The radar itself may do the tracking with servo loops. Alternatively, a track file may be activated in the computer. Depending on the complexity of the tracking technique, the radar may simply look where last seen, move the tracking gate to the nearest target, and record the results. Alternatively, it may develop analog error signals that drive the range gate toward a perfect track. A computer track file usually contains a path model of relatively simple linear form, as given in Equation (3.29).

$$R_3 = R_2 + \left(\frac{R_2 - R_1}{t_2 - t_1} \right) \Delta t \tag{3.29}$$

where:
R_3 = predicted radar-to-target range at time Δt in the future, meters
R_2 = measured radar-to-target range at time t_2, meters
R_1 = measured radar-to-target range at time t_1, meters

As more data are obtained (that is, R_3 is measured and R_4 is predicted), the path may be followed very accurately, especially if it is deterministic (no energy is added to or taken from the object). For indeterminate paths such as an airplane might take, it is important to keep the interval between measurements small as compared to the possible ΔRs during the interval. Sophisticated tracking radars, such as those at Kwajalein Atoll, where the U.S. Department of Defense maintains an extensive missile and reentry vehicle tracking facility, may contain many complex paths in their computers so that, if track is lost, the radar will coast, using the computer programs alone, while looking in designated regions for the target to reappear.

When an object is on a Newtonian trajectory or following some other path that is reasonably deterministic, radar tracking data permit predictions of where that object will be at some future time. The exact equations for these calculations are complex, but they can be approximated by using the so-called "gun barrel analogy" of Figure 3.10. The radar makes measurements over a distance, and the tracker predicts a future target position. Assuming no improvement by integration, the rate of increase of $\Delta\theta$ after the final measurement is given in Equation (3.30). Converting the prediction distance into a velocity-time product and assuming two equal 1σ measurements results in the angular error at the prediction point [Equation (3.31)].

$$\Delta\theta_{dot} = 2\Delta\theta \left(\frac{2L}{l}\right) \tag{3.30}$$

$$\varepsilon_\theta \approx \sqrt{2}\,\Delta\,\theta\left(1 + \frac{2T}{t}\right) \tag{3.31}$$

where:
$\Delta\theta_{dot}$ = Rate of increase in RMS angular error after the final measurement, radians/second
  $\Delta\theta$ = RMS angular error, radians
    L = distance from the last measurement to the prediction point, meters
    l = distance over which measurements are made, meters
  ε_θ = angular error at the prediction point, radians
   T = time ahead the predication is made, seconds
   t = duration of the measurement, seconds

With range errors, the time factor must be introduced initially. Assuming there is no range rate measurement capability, time is measured perfectly, and the range measurements are 1σ, range errors introduce velocity errors propor-

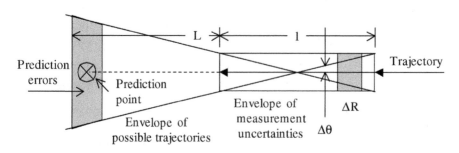

Figure 3.10 Gun-barrel analogy.

tionally, as given in Equation (3.32). Assuming that the first and last range errors are equal and time is measured perfectly, results in the range error at the prediction point are as given in Equation (3.33).

$$\Delta v^2 = \frac{\Delta R_i^2 + \Delta R_f^2}{t} \tag{3.32}$$

$$\varepsilon_R \cong \sqrt{2}\,\Delta R\left(1 + \frac{T}{t}\right) \tag{3.33}$$

where:
Δv = error in the velocity estimate, meters/second
ΔR_i = error in the initial range measurement, meters
ΔR_f = error in the final range measurement, meters
ε_R = range error at the prediction point, meters
ΔR = RMS range error, meters

Many factors serve to modify these elementary results. The trajectory will not be a straight line, and smoothing to a polynomial curve will be necessary. The radar errors will decrease substantially with the number of measurements (N) made, perhaps by as much as $1/\sqrt{N}$. Aspect angle will change during the interval of measurement so that there will be cross correlation between radar range and angle errors. Nevertheless, the simple equations can be used to bound the problem. For example, if the radar has 50-sec measurement duration, a 100-sec prediction time, a 10-m range error, and a 0.5-mrad angular error, we can determine the angle and range errors at the prediction point for a target 900-km from the radar. Equations (3.31) and (3.33) produce a 1σ error volume of ±3182 m cross range and ±42.4 m down range, respectively. This is not far from the numbers obtained by representative radars.

$$\varepsilon_\theta \approx \sqrt{2}(0.5 \times 10^{-3})\left(1 + \frac{2(100)}{50}\right) = 3.5355 \times 10^{-3}\ \text{rad}$$

$$(3.5355 \times 10^{-3})(900 \times 10^3) = 3.182 \times 10^3\ \text{m}$$

$$\varepsilon_R \cong \sqrt{2}(10)\left(1 + \frac{100}{50}\right) = 42.4264\ \text{m}$$

3.6.3 Split Gate Range Tracker

When a single target is being tracked, a split gate range tracker is a very common technique [Stimson, 1998, p. 384; Skolnik, 2001, pp. 246–247]. The range cell is divided into two gates, the early gate and the late gate. A difference signal is generated by the difference of the outputs of the two gates. A closed-loop servo control system is used to drive (center) the split gates over the received target pulse (i.e., zero the difference of the outputs of the two gates).

3.6.4 Leading Edge Range Tracker

If a target tends to be large with respect to a range resolution cell, or if it has scattering phase centers that jump around, "leading edge" and/or "trailing edge" range tracking may be required. The range tracker operates on the pulse rise or pulse fall. A differentiating circuit may be incorporated, as in Figure 3.11, and either or both of the delta functions may be used for the track operations. Because much of the signal power is lost in the differentiating process, high S/Ns are necessary for this type of tracking. The tendency of the range tracker to wander over the "center of mass" of a diffuse target is thereby thwarted.

3.6.5 Doppler Tracking

A Doppler or range rate tracker operates much the same way as the range tracker, except that now the tracking gate is in the frequency domain. A simple

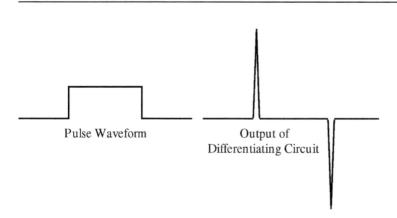

Pulse Waveform

Output of
Differentiating Circuit

Figure 3.11   A differentiating circuit.

effective Doppler tracking system uses banks of narrowband Doppler filters (Doppler filter bank) and circuitry, which provides an indication of which filters are "ringing" (occupied by target Doppler signals). There is also logic to select the most likely range rate band when more than one filter is ringing. These filters are excited only during the instant of the target signal's arrival in the radar, so each response is associated with a given range and angle bin. The revolution in microelectronics has made it feasible to lay thousands of narrowband Doppler filters on substrate along with all the associated logic or to use digital signal processing to implement the fast Fourier transform (FFT). Again, a target that tends to have a diffuse range rate spectrum (spinning propellers, ionized wakes, pieces falling off, and so forth) may require special treatment. A maximum range rate or minimum range rate track may be needed.

3.6.6   Split Gate Doppler Tracker—Speed Gate

When a single target is being tracked, a split gate range rate (or Doppler) tracker is a very common technique [Stimson, 1998, p. 386; Skolnik, 2001, p. 252]. The Doppler cell is divided into two Doppler filters: the low-frequency filter and the high-frequency filter. A difference signal is generated by the difference of the outputs of the two filters. A closed-loop servo control system is used to drive (center) the two Doppler filters over the received target Doppler shift (i.e., zero the difference of the outputs of the two filters).

3.6.7   Target State Trackers: Alpha-Beta and Kalman Filters

Target state trackers are used to enhance the performance of radar systems. They use current target measurements and previous target information to develop a target state estimate, much as we do when we are driving. A target state tracker develops a smoothed, or filtered, estimate of target's state (position, velocity, acceleration, and so forth). It can also develop a predicted target state. Radar tracking concepts and algorithms have evolved rapidly with the exponential evolution of digital computers—speed and capacity increases, and cost and size decrease. The best known of the sophisticated techniques for estimation is the alpha-beta tracker and the Kalman filter, after Kalman and Bucy, who introduced it in 1960. An alpha-beta tracker uses a relatively straightforward set of equations to develop a smoothed target position, smoothed target velocity, and predicted target position, as given in Equation (3.34).

$$\hat{x}_n = x_{pn} + \alpha(x_n - x_{pn})$$

$$\hat{v}_n = \hat{v}_{n-1} + \frac{\beta}{T_s}(x_n - x_{pn})$$

$$x_{p(n+1)} = \hat{x}_n + \hat{v}_n T_s$$

(3.34)

where:

$\hat{x}_n$ = smoothed target position, meters
x_{pn} = predicted target position, meters
α = position smoothing parameter, no units
x_n = measured target position, meters
$\hat{x}_n$ = smoothed target velocity, meters/seconds
β = velocity smoothing parameter, no units
T_s = time between measurements, seconds

If target acceleration is significant, an alpha-beta-gamma tracker is used. Gamma is the acceleration smoothing parameter. There are many ways of selecting the values for alpha and beta. Unfortunately, there are no optimum values. A compromise is made between good smoothing of random measurement errors (narrow bandwidth) and rapid response to maneuvering targets (high bandwidth). An alpha-beta tracker is relatively easy to implement but does not handle maneuvering targets. An alpha-beta tracker can be enhanced with additional algorithms to handle maneuvering targets.

A Kalman filter can inherently handle a maneuvering target through the use of a model of target measurement errors and a model for the target trajectory and its associated uncertainty. The Kalman filter is not a filter at all in the electronic circuit sense. Its solutions to radar estimating problems require extensive and tedious calculations, including matrix inversions, but the method converges rapidly to useful estimators with a minimum of observations. A feature of the Kalman filter is that it operates only in the present, sometimes giving the distillate of all history only equal billing with the latest observation.

A quantitative understanding of the Kalman filter is available from the idea of a "running average" combined with Bayes' formula (published over 200 years ago) for conditional probabilities [Sheats, 1977]. A running average is calculated by subtracting the nth observation from the previous average and dividing it by the number of observations, as given in Equation (3.35). Noting that the running average is an after-the-event measure and that the nth observation is weighted the same as all previous observations, two questions arise. Could the running average be used to estimate future events, and is there an optimum weighting for the most recent observation?

$$\bar{X}_n = \bar{X}_{n-1} + \frac{(O_n - \bar{X}_{n-1})}{n} \tag{3.35}$$

where:
$\bar{X}_n$ = average after n observations
$\bar{X}_{n-1}$ = average after $n-1$
O_n = value of the nth observation

Bayes' formula gives the probability of an event occurring, conditional on another event already having occurred, and it provides for the weighting of the most recent observation based on the quality of that observation compared to the distilled quality of all the previous ones. For radar measurements, the quality of an observation is given in terms of a standard deviation, which is the square root of the variance, and the standard deviation decreases as the number of observations increases, being inversely proportional to the square root of the number of observations. An intuitively appealing weighting would be one that gave heavier weighting to the "best" measurements, "best" in this case meaning those with the smallest variance. A weighting function that has this appeal (it can also be shown to be the optimum) is given in Equation (3.36). Note that, if the variance of the most current measurement is large, the weighting function tends toward zero. If it is small, the weighting function is nearly one.

$$W = \frac{\dfrac{\sigma^2}{n}}{\dfrac{\sigma^2}{n} + \sigma_0^2} \tag{3.36}$$

where:
W = weighting function

$$\frac{\sigma^2}{n} = \text{variance of all measurements that have gone before}$$

$$\sigma_0^2 = \text{variance of the most current measurement}$$

Returning to the equation for a running average, changing the average to an estimate of the mean, and adding the weighting factor results in an equation that illustrates the strategy of the Kalman filter, as given in Equation (3.37). Note how a poor observation, large σ_0^2, will change the estimate little, whereas a good one will be fully effective. As soon as an additional observation is available, the previous estimate becomes part of the running mean, a new weighting function is entered, and a new estimate is calculated.

$$\bar{X}' = \bar{O}_0 + \bar{W}(\bar{X} - \bar{O}_0) \tag{3.37}$$

where:

$\bar{X}'$ = estimate

$\bar{O}_0$ = new observation

$\bar{X}$ = current mean

The Kalman filter estimates are limited to processes that can be written as differential or difference equations. Obviously, discontinuous future events, such as drastic maneuvering by a tracked aircraft, cannot be accommodated. Nevertheless, for predictive tracking problems, and particularly for navigation and guidance applications, the Kalman filter has proven invaluable. An extended Kalman filter adds a model of target dynamics to overcome the limitations of a Kalman filter with respect to drastic target maneuvers. Extended Kalman filters are essential when dealing with missing measurements, variable measurement errors, and maneuvering targets. Extended Kalman filters are widely used in airborne fire control radar systems.

3.7 EXERCISES

1. Use simple detection theory to calculate the S/N required to achieve a single pulse probability of detection $P_d = 0.9$ and probability of false alarm $P_{fa} = 10^{-6}$. Compare with Figure 3.6 or Table 3.1.

2. To increase the performance of a power-limited radar, you plan to use co-herent integration. You currently achieve single-pulse probability of detection $P_d = 0.5$, probability of false alarm $P_{fa} = 10^{-4}$. How many coherent pulses are required to be integrated to obtain $P_d = 0.95$ and $P_{fa} = 10^{-12}$?

3. To make use of the fact that a noncoherent radar's scan pattern illuminates each target with four pulses, a computerized four-pulse detection scheme is proposed. The threshold is set for a probability of detection $P_d = 0.9$ and probability of false alarm $P_{fa} = 0.1$. Calculate the probabilities for P_d and P_{fa} associated with threshold crossing criteria of one, two, three, and four pulses.

4. With single-pulse probability of detection $P_d = 0.5$ and probability of false alarm $P_{fa} = 10^{-4}$, use Figure 3.6 or Table 3.1 to approximate what P_d four pulses of coherent integration would give at $P_{fa} = 10^{-8}$.

5. Looking close to the horizon at low-flying satellites, a radar can see range rates of 4000 m/sec at ranges of about 2000 km. Visualize a satellite catalog-ing radar with a beamwidth of 1 degree, required to cover 50 × 2 beam-widths with two pulses each at a maximum range of 4000 km. Assuming no special beam forming and 1 msec for beam switching, how many tracks at the same range could this radar accomplish with a 1-sec data rate? With a 0.20-sec data rate? Hints: A double layer of beamwidths is interleaved so that it is 1.866 beamwidths thick. Calculate satellite transit time through the stack (13-sec), round-trip time to 4000 km (0.024-sec), proportion of time left for tracks (8/13).

6. A radar uses a five-attempt binomial detection with a three-out-of-five threshold-crossing criterion. The single attempt probability of detection $P_d = 0.5$, probability of false alarm $P_{fa} = 10^{-4}$, and associated detection threshold S/N = 9.4 dB (Figure 3.6 or Table 3.1). What is the probability three or more detection attempts? What is the probability of false alarm as-sociated with three or more threshold crossings? Use Figure 3.6 or Table 3.1 to determine the detection threshold associated with the probability of de-tection and probability of false alarm after three or more detection at-tempts.

7. A radar has a probability of detection $P_d = 0.4$, and probability of false alarm $P_{fa} = 10^{-6}$ for each individual detection event. Calculate the cumulative probabilities of detection and false alarm when six individual detections are combined.

8. When the radar in Exercise 6 uses an at least three-out-of-five threshold crossing detection criterion, its detection range is $R_d = 100$ km. What is the detection range if the radar uses a one threshold crossing detection criterion? [Hint: Use the results of Exercise 6.]

3.8 REFERENCES

Barton, D. K., 1974, *Radars*, Vol. 1, "Monopulse Radars," Norwood, MA: Artech House. This volume is a collection of reprints of papers on monopulse radar.

Barton, D. K., 1979, *Radar System Analysis*, Norwood, MA: Artech House. Abstruse theory is made understandable in this detailed work, oriented toward analysis and test of radar systems.

Barton, D. K., 1988, *Modern Radar System Analysis*, Norwood, MA: Artech House.

DiFranco, J. V., and Rubin, W. L., 1980, *Radar Detection*, Norwood, MA: Artech House. DiFranco and Rubin's book is the standard for radar detection; however, be prepared for the calculus. They are meticulous in their development, solution, and presentation of the results.

Freund, J. E., 1962, *Mathematical Statistics*, Englewood Cliffs, NJ: Prentice Hall. This book is not for someone who wishes to teach himself mathematical statistics.

Marcum, J. T., 1960, "A Statistical Theory of Target Detection by Pulsed Radar," *IRE Transactions in Information Theory*, Vol. IT-6, Apr., pp. 145–267. Marcum's work has stood the test of time.

North, D. O., 1943, "An Analysis of Factors Which Determine Signal/Noise Discrimination in Pulsed Carrier Systems," RCA Labs Tech Report PTR6C (reprinted in *Proceedings of the IRE*, Vol. 51, July 1963, pp. 1016–1027).

Rhodes, D. R., 1959, *Introduction to Monopulse*, New York: McGraw-Hill.

Schwartz, M., 1959, *Information Transmission, Modulation, and Noise*, New York: McGraw-Hill.

Sheats, L., 1977, "The Kalman Filter," Chap. 26 in E. Brookner (Ed.), *Radar Technology*, Norwood, MA: Artech House.

Skolnik, Merrill I., 2001, *Introduction To Radar Systems*, 3rd ed., New York: McGraw-Hill. Skolnik is meticulous about presenting all the equations and all the references. He emphasizes answers.

Stimson, George W., 1998, *Introduction to Airborne Radar*, 2nd ed., Raleigh, NC: SciTech Publishing.

4

Radar Cross Section

Radar cross section (RCS) is a measure of the electromagnetic energy intercepted and reradiated at the same wavelength by any object. The dimensions are those of an area, usually square meters (m^2) or decibels relative to a square meter (dBsm). The relationship between square meters and dBsm is given in Equation (4.1). The RCS of an object is a complex combination of multiple factors: size, shape, material, edges, wavelength, and polarization. Simple objects tend to have a single, or few, scattering sources. Complex objects (such as airplanes) tend to have multiple scattering sources (e.g., nose, fuselage, inlet, wing root, wing, and so forth). Thus, for complex objects, the RCS is the complex (amplitude and phase) combination of contributions from each scattering source.

The understanding of RCS concepts generally starts with an idealized object that is large with respect to a wavelength, has an intercept area of one square unit, is perfectly conducting, and reradiates isotropically. It is easy to build an object with these characteristics; a copper sphere is an example. Providing it is large with respect to the wavelength of the incident electromagnetic energy, a copper sphere of projected area of 1 m^2 has radar cross section, usually indicated by σ, of 1 m^2 or, for any sphere of radius a, $2\pi a/\lambda > 10$, as given in Equation (4.2).

$$\text{dBsm} = 10 \log(\text{m}^2) \qquad \text{m}^2 = 10^{\left(\frac{\text{dBsm}}{10}\right)} \qquad (4.1)$$

$$\sigma_s = \pi a^2 \qquad (4.2)$$

where:
  σ_s = radar cross section of sphere, $2\pi a/\lambda > 10$, square meters
  a = radius of the sphere, meters
  λ = wavelength, meters

The RCS of an object can be determined by solving Maxwell's equations. The RCS of an object can also be established by measuring it and comparing it to that of the reference object. For complex, many-surfaced objects (such as airplanes), attempts to solve Maxwell's equations for the various boundary conditions have not been very successful. Measurement and comparison then became the only way of obtaining RCS. As the characteristics of the object vary with aspect angle, so does the RCS, often fluctuating rapidly. Computer programs are able to calculate the radar cross section for some rather complicated bodies. Computer programs essentially break down a complicated target into many simple surfaces and superpose their radar cross sections (amplitude and phase) to compute the overall radar cross section.

For several classes of simple objects, RCS can be easily calculated. When the target does not reradiate isotropically, it may have gain in the direction of the radar, as given in Equation (4.3).

$$\sigma = G A_e \qquad (4.3)$$

where:

σ = radar cross section for nonisoptropic object, square meters
G = reradiation gain in the direction of the radar, no units
A_e = electromagnetic area of the object as seen by the radar, square meters

In Chapter 2, we showed from first principles that the reradiation gain is a function of the electromagnetic area and wavelength [Equation (4.4)]. It should always be remembered that the area is an electromagnetic area as seen by the radar at the instant in time that radar cross section is being induced. From this, the calculation of the RCS of a flat plate is trivial [Equation (4.5)].

$$G = \frac{4\pi A_e}{\lambda^2} \qquad (4.4)$$

$$\sigma = \frac{4\pi A_e^2}{\lambda^2} \qquad (4.5)$$

The calculation of such retrodirective targets as corner reflectors where the radar is seeing the equivalent of a flat plate is straightforward. Figure 4.1 is a sketch of a two-face (dihedral) and a three-face (trihedral) corner reflector. The faces are at right angles to each other. Some basic trigonometry reveals that many plane waves arriving at these reflectors are directed back toward the

Dihedral Trihedral

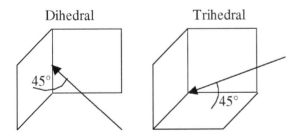

45°

45°

Note: Arrows indicated entry angle for maximum RCS

Figure 4.1 Two corner reflector designs.

radiating source in phase, making the maximum RCS equal to the projected area of the corner reflector [Knott, Schaeffer, and Tuley, 1985, p. 178]. If all facets of these reflectors are equal sized squares, then the RCS of the dihedral and trihedral are given in Equations (4.6) and (4.7), respectively.

$$\sigma_d = \frac{8\pi \, a^4}{\lambda^2} \qquad\qquad (4.6)$$

$$\sigma_t = \frac{12\pi \, a^4}{\lambda^2} \qquad\qquad (4.7)$$

where:
  σ_d = radar cross section of a dihedral corner reflector, square meters
  a = dimension of the square, meters
  σ_t = radar cross section of a trihedral corner reflector, square meters

4.1 RCS OF A SPHERE

A calculation can also be made to determine the amount of a sphere that is sufficiently flat that it scatters back to the radar. From the ratio of its RCS to its physical surface area, we have already established that the RCS is one-fourth the surface area. This can also be derived using the geometry of Figure 4.2, as given

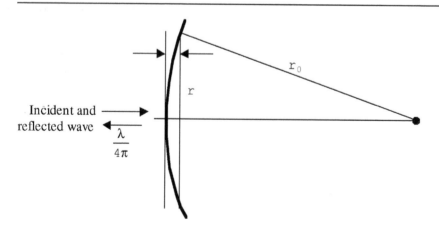

Figure 4.2 Geometry for sphere RCS.

in Equations (4.8) and (4.9). The result is reasonably useful, because it permits the calculation of RCS for large, curved, irregular bodies based on the radii of curvature of their convex surfaces. The RCS of any one surface is given in Equation (4.10).

$$r \cong \sqrt{2r_0 \frac{\lambda}{4\pi}} \qquad A = \pi r^2 = \pi 2 r_0 \frac{\lambda}{4\pi} \tag{4.8}$$

$$\sigma_s = \frac{4\pi A^2}{\lambda^2} = \pi r_0^2 \tag{4.9}$$

where:
   r_0 = radius of the sphere, meters
   σ_s = radar cross section of a sphere, square meters

$$\sigma_c = \pi r_x r_y \tag{4.10}$$

where:
   σ_c = radar cross section of a large curved convex shape, square meters
   r_x = radius of curvature in the x-plane, meters
   r_y = radius of curvature in the y-plane, meters

These discussions of RCS result in two important, generally applicable pieces of information on radar cross section. First, the average RCS of large and irregular but smooth objects is equal to one-quarter their surface area. Second, the RCS of multifaceted objects with convex surfaces facing the radar can be calculated.

For objects with at least one dimension that is small with respect to a wavelength, the simple relationships no longer hold. Again, the conducting sphere is an excellent example. When the circumference of the sphere becomes about equal to a wavelength, the currents induced on the surface of the sphere radiate and add to the reflected wave. When the circumference, $2\pi r$, is somewhat greater than $3\lambda/2$, the radiation from induced currents tends to cancel the direct radiation. At $2\pi r > 2\lambda$, there is enhancement again; at $2\pi r > 5\lambda/2$, there is cancellation, and so forth. This zone of enhancement and cancellation is called the *Mie* or *resonance region*. As the circumference becomes several λ, there are bands of current traveling in opposite directions on the body so that the effects of addition and subtraction are damped. The result of all these phase additions and subtractions is the famous curve for the RCS of a conducting sphere shown in Figure 4.3 [Crispin and Maffett, 1968, pp. 86–87; Rheinstein, 1968, p. 5]. Although this curve is strictly applicable only to spheres, it is approximate for many bodies with roughly equal measurements in all three dimensions, and, in the Rayleigh region, it becomes a fairly accurate measurer of the volume of many objects. The Rayleigh region was defined in classical physics. Lord Rayleigh first worked out the λ^4 dependency of the amount of light scattered by molecules in the Earth's atmosphere. The relationship is analogous to the scattering of electromagnetic waves from objects that are small compared to wavelength. Although straightforward, the derivation is lengthy. It is included in many university texts of basic optics [Jenkins and White, 1976, pp. 467, 471–473]. RCS in the Rayleigh region is given in Equation (4.11).

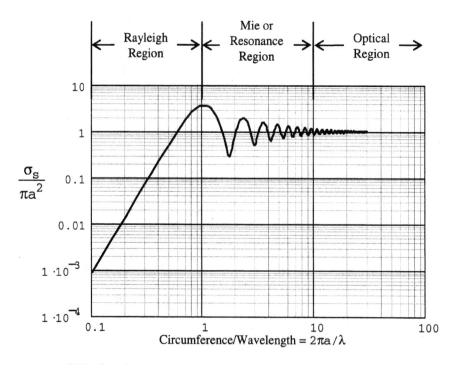

Figure 4.3 RCS of a sphere.

$$\sigma_{Rayleigh} = 8\pi^3 \frac{V^2}{\lambda^4} \tag{4.11}$$

where:

$\sigma_{Rayleigh}$ = radar cross section of a sphere in the Rayleigh region, square meters
V = volume of the sphere, cubic meters (m^3)

4.2 RCS OF SIMPLE OBJECTS

From the idea of intercept area and reradiated gain, the broadside RCS of a cylinder is easily deduced. A cylinder has projected area of the product of its length and diameter. It reradiates isotropically in the plane perpendicular to its axis of symmetry. Its RCS is given in Equation (4.12). End-on, it is a flat plate.

$$\sigma_{cly} = \frac{\pi L^2 d}{\lambda} \tag{4.12}$$

where:

σ_{cly} = radar cross section of a cylinder, square meters
L = length of the cylinder, meters
d = diameter of the cylinder, meters

If we let the diameter get small, we have a long wire. Electrically, the diameter does not become vanishingly small, however; its contribution decays exponentially with the ratio of the wire diameter to the wavelength. Crispin and Maffett [1968] worked out a relationship that applies to wires many wavelengths long with diameters of two wavelengths or less, as given in Equation (4.13).

$$\sigma_w = \pi L^2 \left(\frac{d}{2\lambda}\right)^{0.57} \tag{4.13}$$

where:

σ_w = radar cross section of a long wire, square meters

The radar cross sections of objects with characteristic dimension $2a$ tend to be like πa^2 as long as $2\pi a \sim \lambda$. So it is that reentry vehicles (RVs) with base diameters of about 0.6 m tend to have at VHF frequencies, where λ = 1.8 m, a nose-on RCS of approximately 1 m^2. At higher frequencies, where geometric optics prevail and the induced currents cancel or are attenuated, only the nose tip and discontinuities such as joins cause reflections back to the radar from a nose-on aspect. A smooth ice-cream-cone RV will tend to have $\sigma_{ave} = 0.1\lambda^2$ until nose tip scattering becomes a factor. The RCS of the nose tip is given in Equation (4.14).

$$\sigma_{nose\ tip} = \pi 7 a_{nt}^2 \qquad a_{nt} \geq \frac{\lambda}{2\pi} \qquad\qquad (4.14)$$

where:

$\sigma_{nose\ tip}$ = radar cross section of nose tip, square meters
a_{nt} = radius of the nose tip, meters

For a 25-mm nose radius at S-band, $\sigma = 0.002\ m^2 = -27$ dBsm. An RV with a flat back will have an RCS varying between the sum and difference of the $\sigma_{nose\ tip}$ and the $\sigma_{flat\ back}$. The latter approximates the RCS of a wire loop of the same circumference and therefore is the dominant scatterer, so its RCS is given in Equation (4.15). The RCS of these and other simple, perfectly conducting shapes are shown in Table 4.1. Many of these shapes will be discussed in more detail in the following sections.

$$\sigma_{flat\ back} = \pi r^2 \qquad r \geq \frac{\lambda}{2\pi} \qquad\qquad (4.15)$$

where:

$\sigma_{flat\ back}$ = radar cross section of the flat back, square meters
r = radius of the flat back, meters

4.3 POLARIZATION

Thus far, we have said nothing about the effect of polarization on RCS. The effects of polarization on RCS can be dramatic. Smooth targets do not depolarize incident waves; therefore, the return wave has the same polarization as the incident wave but in the opposite sense (that is, the electric vector has flipped over). A weather radar designer who wanted to view droplets and clouds would not worry excessively about polarization. However, a radar meant to view complex (and depolarizing) targets through weather might transmit horizontal po-

Table 4.1   Radar Cross Sections of Simple Shapes

Shape	Aspect Angle	RCS	Constraints	Symbols
	Any	πr^2	$\dfrac{2\pi r}{\lambda} > 10$	r = radius
Sphere	Any	$8\pi^3\left(\dfrac{V^2}{\lambda^4}\right)$	$\dfrac{2\pi r}{\lambda} < 1$	V = volume
Flat plate	Broadside	$\dfrac{4\pi A^2}{\lambda^2}$	$\dfrac{2\pi a}{\lambda} > 1$	A = area a = dimension
Cone	Axial	$\dfrac{\lambda^2}{16\pi}\tan^4(\theta)$		θ = cone half angle
Cone sphere	Nose-on	$0.1\lambda^2$	$\dfrac{2\pi r_{\text{nose tip}}}{\lambda} < 1$	$r_{\text{nose tip}}$ = nose tip radius
		$\approx \pi r_{\text{nose tip}}^2$	$\dfrac{2\pi r_{\text{nose tip}}}{\lambda} > 1$	
Truncated cone	Nose-on	$\approx \pi r_{\text{flat back}}^2$ $\pm\ \sigma_{\text{nose tip}}$	$\dfrac{2\pi r_{\text{flat back}}}{\lambda} > 1$	$r_{\text{flat back}}$ = flat back radius
Cylinder	Broadside	$\dfrac{\pi d L^2}{\lambda}$	$\dfrac{2\pi d}{\lambda} > 1$	d = diameter L = length
Long wire	Broadside	$\pi L^2\left(\dfrac{d}{2\lambda}\right)^{0.57}$	$\dfrac{L}{\lambda} \gg 1$	d = diameter L = length
Resonant dipole	Broadside	$\dfrac{\pi}{4}\lambda^2$	$L = \dfrac{\lambda}{2}$	L = length
	Spherically random	$0.15\lambda^2$		
Cloud of resonant dipoles	Random	$0.15\ N\lambda^2$	$L = \dfrac{\lambda}{2}$	N = number of dipoles
Convex surfaces	Random	$\pi r_x r_y$	$(\lambda < \ < r_x)$ and r_y	r_x = radius of curvature x-plane r_y = radius of curvature, y-plane
Dihedral corner reflector	Axis of	$\dfrac{8\pi a^4}{\lambda^2}$	$\lambda < a$	a = dimension
Trihedral corner reflector	Axis of	$\dfrac{12\pi a^4}{\lambda^2}$	$\lambda < a$	a = dimension
Triangular corner reflector	Axis of	$\dfrac{4\pi a^4}{3\lambda^2}$		a = dimension

larization and receive vertical. Or (as is usual) the radar might transmit right circular and receive left circular polarization. Thus, raindrop returns would be filtered out. Aircraft control radars employ this technique.

In general, the effects of polarization can be estimated by knowing the angles made with the incident electric vector by edges and long, thin pieces of the target, as given in Equation (4.16). The average RCS is given in Equation (4.17). This result says that, on average, a randomly depolarizing target returns only half the power of a nondepolarizing target. Of course, the target may have preferential polarization angles that would greatly reduce the return. When the target's polarization characteristics are unknown, the return of at least $\sigma_0/2$ can be assured by transmitting circular polarization.

$$\sigma = \sigma_0 \cos^2(\theta) \tag{4.16}$$

$$\bar{\sigma} = \frac{2\sigma_0}{\pi} \int_0^{\pi/2} \cos^2(\theta)\, d\theta = \frac{\sigma_0}{2} \tag{4.17}$$

where:
  σ = radar cross section at a polarization angle, square meters
  σ_0 = radar cross section of a non-depolarizing target, square meters
  θ = polarization angle, radians
  $\bar{\sigma}$ = average radar cross section across all polarization angles, square meters

The return signal is further modified by the characteristics of the receive system. It may receive the same polarization it transmits, or it may receive on two orthogonal polarizations. It is easy to speak in quantitative terms about the various alternatives.

If the intervening medium rotates the polarization angle in a random way, a horizontally polarized transmitter, a randomly depolarized target, and a horizontally polarized receiver will deliver an average RCS as given in Equation (4.18). If the intervening medium does not further depolarize or rotate the polarization angle, then the average RCS is given in Equation (4.19).

$$\bar{\sigma}_{HH} = \frac{2\sigma_0}{\pi} \int_0^{\pi/2\pi} \cos^2(\theta)\, d\theta \cdot \frac{2}{\pi} \int_0^{\pi/2} \cos^2(\theta)\, d\phi = \frac{\sigma_0}{4} \tag{4.18}$$

$$\bar{\sigma}_{HH} = \frac{2\sigma_0}{\pi} \int_0^{\pi/2} \cos^4(\theta)\,d\theta = \frac{3\sigma_0}{8} \tag{4.19}$$

where:

$\bar{\sigma}_{HH}$= average radar cross section (horizontal transmit–horizontal receive), square meters

ϕ = random polarization angle, radians

If a vertically polarized receiver is added, the average RCS is given in Equation (4.20). The sum of the power in the two (horizontal and vertical) channels is $\sigma_0/2$, which is all that is available from the target. Thus, a radar that deals with targets and media having largely uncertain polarization characteristics minimizes the possible impact of polarization effects by transmitting circular polarization and receiving on two orthogonal linear polarizations. This arrangement guarantees that, if the receive channels are summed, the average RCS equals $\sigma_0/2$.

$$\bar{\sigma}_{HV} = \frac{2\sigma_0}{\pi} \int_0^{\pi/2} \cos^2(\theta)\sin^2(\theta)\,d\theta = \frac{\sigma_0}{8} \tag{4.20}$$

where:

$\bar{\sigma}_{HV}$ = average radar cross section (horizontal transmit—vertical receive), square meters

4.4 CHAFF CHARACTERISTICS

One class of targets whose performance is greatly polarization sensitive is chaff. Chaff is small, light pieces of material that have high radar cross section. The use of chaff as an electronic countermeasure is discussed in Chapter 6. The RCS of a single half-wave resonant dipole can be calculated analytically in several ways [Van Vleck, Block, and Hammermesh, 1947]. Experimental results, however, do not agree absolutely with the detailed analysis. It is therefore more convenient to use simple theory to derive radar cross section, especially because the experimental measurements agree relatively closely with simple theory.

Returning to the ideas developed in Chapter 2, we can find the maximum gain of a half-wave dipole in free space. Recall that when the far field of an array contains no grating lobes, its near field is fully filled. This occurs when the element separation does not exceed $\lambda/2$ and implies that the intercept area of an element in a fully filled array cannot exceed $\lambda^2/4$. Applying the equation $G = 4\pi A/\lambda^2$ to a dipole element gives $G = \pi$. However, this is a dipole backed by a conducting surface, doubling the gain. Radiation into free space reduces G to $\pi/2$. The same result can also be arrived at by noting that the dipole pattern is like a donut with an infinitesimal hole. Around its axis, the gain is one. Across its axis (as with the cylinder), $G = \pi L/\lambda$. For the half-wave dipole, $L = \lambda/2$, making $G = \pi/2$.

The peak RCS of a dipole is given as $0.88\lambda^2$ [Barton, 1988, p. 103]. But that is a peak value, not an average within the dipole's defined beamwidth. If the effective dimensions of the intercept area of a dipole in free space are assumed to be enhanced to λ along the dipole axis and $\lambda/2$ across it, making $A = \lambda^2/2$, we can apply $\sigma = GA$ to get Equation (4.21) (which is close to Barton's value).

$$\sigma_p = \left(\frac{\pi}{2}\right)\left(\frac{\lambda^2}{2}\right) = \frac{\pi\lambda^2}{4} \approx 0.8\lambda^2 \tag{4.21}$$

where:

σ_p = peak chaff radar cross section, square meters

The average RCS of a half-wave dipole can now be found by assuming an spherically random distribution, applying the two-way $\sin\theta$ amplitude effects ($\sin^2\theta$ power) and $\cos^2\phi$ polarization. The average chaff RCS is given in Equation (4.22). The integration is available from integral tables, and the result is given in Equation (4.23). Others have used from $0.13\lambda^2$ to $0.2\lambda^2$ for this value.

$$\bar{\sigma} = \sigma_p \frac{4}{\pi^2} \int_0^{\pi/2} \sin^4(\theta) \cos^2(\phi)\, d\theta\, d\phi \tag{4.22}$$

$$\bar{\sigma} = \frac{3\sigma}{16} \approx 0.15\lambda^2 \tag{4.23}$$

where:

$\bar{\sigma}$ = average chaff radar cross section, square meters

It is now a simple matter to determine the peak and average RCS of a cloud of randomly oriented, perfectly conducting dipoles, as given in Equations (4.24) and (4.25), respectively. When dealing with radar results, some care must be taken to differentiate between the radar resolution volume and the cloud volume. If the former is smaller than the latter in any dimension, the RCS per cell must be summed to find the cloud RCS.

$$\sigma_p \cong 0.8 N \lambda^2 \tag{4.24}$$

$$\bar{\sigma} \cong 0.15 N \lambda^2 \tag{4.25}$$

where:
   N = number of chaff dipoles in the cloud, no units

The number of chaff particles in a radar resolution cell has a distinct effect on how the return is viewed by the radar. The RCS of several chaff particles in a single radar resolution cell is noise-like and can be treated like noise for the purposes of detection and signal-processing theory. The RCS of numerous chaff particles in a single radar resolution cell is a target-like return and can be treated like a target for the purposes of detection and signal-processing theory.

4.5 DIFFUSE TARGETS AND CLUTTER

Chaff RCS represents the scattering from a diffuse target. Scattering from various types of diffuse targets is an important topic for discussion. Radar altimeters and synthetic aperture (imaging) radars, as examples, make extensive use of diffuse scatterers, such as multiple objects on the ground and the ground itself. All the considerations about scattering we have touched on so far are applicable when developing scattering models for various types of ground targets. Wavelength dependence is strong. The leaves of trees, for example, may have large, noisy RCS at X-band ($\lambda \approx 0.03$ m) but be virtually invisible at UHF where $\lambda \approx 0.6$ m. The ground itself tends to be diffuse except in areas of rocky terrain (where objects encountered tend to be large with respect to wavelength) and in sandy regions (where the roughness scale is much, much smaller than a wavelength). Calm water will be a specular reflector whose radar cross section is very sensitive to the incidence angle. Rough water may be diffuse or specular, and periodic, depending on the scale size and makeup of the waves. Man-made objects tend to be specular.

A number of investigators have gathered a great deal of detailed data on clutter [Barton, 1975; Long, 2001; Billingsley, 2002]. Such studies, which are absolutely essential for design of a particular radar for a specific mission, tend to obscure important general principles. Sea and ground clutter are usually treated separately. Although there are many similarities in clutter effects, the excursions of sea clutter are more extreme. To facilitate incorporation into the radar equation, the clutter reflectivity (radar cross section per unit area) is usually used in the literature, as given in Equation (4.26). Example clutter reflectivity and reflection coefficient values are shown in Table 4.2 [Stimson, 1998]. Extensive tables and figures of clutter reflectivity and/or reflection coefficients are given in Long (2001) and Billingsley (2002).

$$\sigma_0 = \Psi \sin(\alpha) \qquad (4.26)$$

where:

σ_0 = clutter reflectivity, m^2/m^2

Ψ = reflection coefficient, m^2/m^2

α = grazing angle of the radar wave relative to the ground, radians or degrees

Table 4.2   Typical Backscattering Coefficients

Terrain	Clutter Reflectivity (dBsm)	Reflection Coefficient (dBsm)
Smooth water	−53	−45.4
Desert	−20	−12.4
Wooded area	−15	−7.4
City	−7	0.6
	10° grazing angle and 10 GHz	

When the grazing angle is between, say, 10° and 60° (that is, the radar is looking neither at the horizon nor straight down), for many classes of terrain and wavelengths, grazing angle dependence is small and may not be used. The reason is that, for smooth surfaces (granules $\ll \lambda$), there will be virtually no clutter return up to very high angles of incidence; for rough surfaces (granules $\leq \lambda$, whose individual radar cross sections are not very dependent on angle), the clutter return will tend to be constant in the region between a few degrees above $0°$ and a few degrees below $90°$. Barton (1985, pp. 198–204) has noted the lack of a model for low grazing angles and provided one that includes both reflectivity and propagation characteristics.

Of the models of radar cross section developed for various simple objects in the preceding sections, some depend on λ^2, some depend on λ, and some are wavelength independent. In the Rayleigh region, they are λ^4 dependent, but their absolute level tends to be sufficiently low as to avoid being a factor. One could model terrain by using various combinations of plates, spheres, cylinders, and wires, making them conducting and nonconducting, as appropriate. It is common to find such a model dependent on λ, which the data confirm. Also to be expected is for ψ to vary over several orders of magnitude, and it does. For water, it may be between 0.1 and 0.001; for flat, desert-like country, 0.01; and 0.1 for dense undergrowth. The fact that even water is not a very good conductor makes the returns lossy; so ψ does not rise near 1.0 except at very high incidence angles, such as with radar altimeters, which look straight down. Here, ψ may exceed 10 over smooth water. In a super simplification of a large body of data on land clutter (arguing that dependence on λ and α are both small), Nathanson (1969, p. 272) calculates the rough equivalent of our ψ to be between 0.5 and 0.001 with a median at 0.04.

Wind is a factor in determining ψ, especially at sea. Because clutter rejection filters need to be narrow, clutter range rate spread is very important in assessing the efficiency of clutter rejection techniques (see Chapter 7, "Moving Target Indication").

For ground-based radars attempting to detect and track airborne targets, ground targets and the ground itself are classified as clutter. This clutter must be rejected in some way. Clutter fences around the site are an efficient way to do this for some clutter problems. Clutter fences merely prevent the radar from illuminating relatively distant objects (such as mountains) whose enormous RCS is at the range where targets must be observed. Instead, the energy reflects from the clutter fence at a range so short that the unwanted reflections are easily rejected. Because ground clutter is in the sidelobes of the radar beam, and sidelobe level tends to be inversely proportional to frequency squared (for a constant aperture size), clutter fences are usually mandatory only in the lower microwave regions. For near-in clutter, the radar receivers are simply not turned on until after the outgoing pulse has returned from clutter. Where there are points of very high clutter returns (say, buildings, walls, or roads), in the radar coverage, these may be blanked in the radar receiver, but at the cost of blanking targets at that range and angle.

When clutter is at the same range and angle as the target, the problem of rejecting it is much more difficult. Moving-target-indicator and pulse-Doppler processing were invented to assist in solving this problem. These techniques are discussed in ensuing chapters—theory in Chapter 5 and applications in Chapter 7.

In the special case of imaging radars, the clutter literally becomes the target and is the major portion of the data gathered. Radar mapping is also covered in Chapter 7.

For radars operating against diffuse clutter, the range equation is modified. The clutter in which the target is embedded is noise-like but at much greater amplitude than the receiver thermal noise. This type of clutter is often called *competing clutter*, as it is in the same resolution cell as the target, and thus the target competes with the clutter for detection. The receiver thermal noise, therefore, drops out of the range equation, and an expression for the competing clutter power is substituted. The competing signal-to-clutter ratio (S/C) is determined where energy is simultaneously scattered back from both the target and the clutter. The RCS of the clutter is simply the clutter reflectivity multiplied by the area of clutter intercepted by the radar pulse, as shown in Equation (4.27). The area of the clutter intercepted by the radar pulse is the product of the projection of the cross-range and range resolutions (i.e., resolution cell size) on the ground, as shown in Equation (4.28). The return from the target is simply the target's RCS. The S/C is the ratio of the signal power, Equation (1.11), to the clutter power, Equation (1.11), using the RCS of the clutter, as shown in Equation (4.29). An exercise at the end of the chapter asks the reader to derive the S/C equation.

$$\sigma_c = \sigma_0 A_c \tag{4.27}$$

$$A_c = (R\theta_{3dB})\left(\frac{c\tau}{2}\right)\left(\frac{1}{\cos(\alpha)}\right) \tag{4.28}$$

$$\frac{S}{C} = \frac{2\cos(\alpha)\,\sigma}{\sigma_0\,R\,\theta_{3dB}\,c\,\tau}$$

(4.29)

where:

σ_c = clutter radar cross section, square meters

A_c = area of the clutter intercepted by the radar pulse, square meters

R = radar-to-target (and clutter) range, meters

θ_{3dB} = radar antenna azimuth 3-dB beamwidth, radians

c = speed of light, 3×10^8 meters/second

τ = pulse width, seconds

S/C = signal-to-clutter ratio, no units

σ = target radar cross section, square meters

If the clutter returns are high, which they usually are, no amount of increased radar power will redress the situation. Note, too, that S/C becomes worse as the range increases. A radar in this condition is said to be *clutter-limited;* its detection range is limited by clutter instead of receiver thermal noise. As one of the exercises demonstrates, a radar that has not been designed for the task simply cannot detect targets in clutter. Targets below airborne radars will be obscured, and low-flying targets may be undetectable by ground-based radars until ranges are sufficiently long to place the clutter below the horizon. Two ways of improving the S/C are apparent from the equation: make the radar transmit beamwidth narrower, and make the transmit pulses shorter. There is a third approach: separate the target from the clutter by the Doppler (range rate) differences between them, a subject to be taken up in Chapter 5. The above discussion ignores sidelobe clutter at the same range as the target. In radars with good (low) sidelobes, the additional effect of clutter in all the sidelobes can be minimized. In radar with poor (high) sidelobes, the additional effect of clutter in the sidelobes should be considered.

4.6 RADAR SIGNATURES

We have alluded to the antenna-like characteristics of various objects whose RCS we have discussed. We have also speculated about the physical features that result in certain RCS behavior. The next step is to ask what might be determined about objects' physical characteristics by studying their radar cross sections. This is the field of *radar signatures*. It is substantial, and it goes beyond the study of far-field RCS patterns. Nevertheless, some things can be learned about an object from observing its far-field patterns rolling past a radar's line of sight.

Referring back to the elemental antenna of Chapter 2, we can make it equivalent to an RCS far field by noting that the energy sources arrive from the radar. The effect is that the pattern changes twice as fast as we move away from perpendicular, making the lobes half as wide. The peak-to-first-null width is now λ/2D instead of λ/D.

Figure 4.4 shows the RCS patterns of some simple objects rotated through 180°. The inferences that might be made as to target characteristics are obvious [Crispin and Maffett, 1968, pp. 83–153]. If we can find out the angular rate at which the objects are turning, we can get a measurement of their width. The

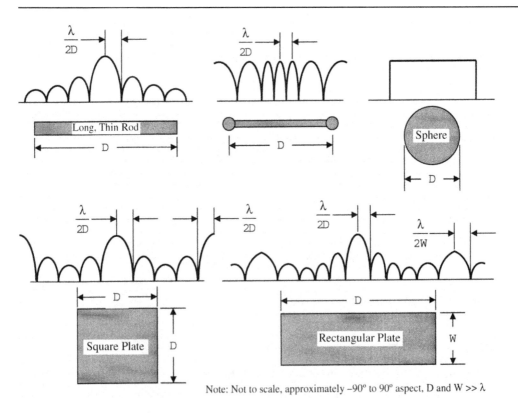

Note: Not to scale, approximately –90° to 90° aspect, D and W >> λ

Figure 4.4 Some simple RCS patterns.

"beamwidth" of the lobe in the RCS pattern of a rotating object is given in Equation (4.30).

$$\theta = 2\pi\left(\frac{\Delta t}{T}\right) = \frac{\lambda}{2D} \Rightarrow D = \frac{\lambda}{4\pi}\left(\frac{T}{\Delta t}\right) \tag{4.30}$$

where:

θ = peak to first null in the RCS pattern, radians

Δt = time it takes for the lobe in the RCS pattern to cross the radar line of sight, seconds

T = time it takes for the object to rotate one time, seconds

λ = wavelength, meters

D = dimension of the object, meters

Much can be done with this kind of analysis using reasonable assumptions, particularly with space objects, where the motion is Keplerian and the observation periods can be long. However, we must always remember that there is not a one-to-one correspondence between an illumination function and its far field. In addition, there most emphatically is not a one-to-one correspondence between an illumination function and the external physical characteristics of an object. Finally, the external characteristics of the object may tell us nothing of its internal characteristics.

4.7 EXERCISES

1. You are estimating the RCS of a particular type of aircraft that will frequently fly through the coverage of a surveillance radar you are designing. You find that it has an L-band antenna of area, A, that will be directly in the radar line of sight much of the time. Your radar is at X-band. Assume the antenna to be a flat plate.

   A. What will the RCS of that antenna be?

   B. If it were 50% efficient and matched to your radar, what would its RCS be?

2. You are supervising the designers of a radar for finding and mapping rainfall. In the region where it will operate, raindrops range in diameter from 2.5 to 13 mm. What frequencies should be used?

3. An aircraft of 30-m wingspan is flying toward a radar of frequency f_c = 900 MHz. Over what changes in aspect angle would the aircraft have to move to make a measurement of wingspan?

4. Using the tenets of antenna theory developed in Chapter 2 and RCS principles discussed in this chapter, derive the approximate RCSs for a sphere, dipole, flat rectangular plate, long wire, and cylinder. (Use $\sigma = GA$, and $G = \dfrac{4\pi A}{\lambda^2}$.) See Table 4.1.

5. 100,000 aluminum chaff dipoles might weigh a few pounds at L-band (λ = 0.3 m). What would be the cloud's average RCS?

6. What would be the signal-to-clutter ratio for a radar with a 2° beamwidth antenna and a 1-μs pulse trying to see a 1 m^2 RCS target at 10-km range in clutter having a σ_0 of –30 dBsm and a 1-degree grazing angle?

7. Derive the equation for signal-to-clutter ratio (S/C) [Equation (4.29)].

4.8   REFERENCES

Barton, D. K., 1975, *Radars*, Vol. 5, *Radar Clutter*, Norwood, MA: Artech House. Barton has gathered 38 articles on the subject of land, sea, and aerospace clutter.

Barton, D. K., 1985, "Land Clutter Models for Radar Design and Analysis," *Proc. IEEE*, Feb. 1985, pp. 198–204.

Barton, D. K., 1988, *Modern Radar System Analysis*, Norwood, MA: Artech House.

Billingsley, J. B., 2002, *Low-Angle Radar Land Clutter Measurements and Empirical Models*, Raleigh, NC: SciTech Publishing. Just about anything one would want to know about land clutter.

Crispin, J. W., and A. L. Maffett, "RCS Calculation of Simple Shapes-Monostatic," in J. W. Crispin and K. M. Siegel (Eds.), *Methods of Radar Cross Section Analysis*, 1968, New York: Academic Press. Extensive theory and detailed calculations for many shapes of varying orientations and polarization angles are provided.

Jenkins, F. A., and H. E. White, 1976, *Fundamentals of Optics*, New York: McGraw-Hill.

Knott, F. F., J. F. Schaeffer, and M. T. Turley, *Radar Cross Section*, 1985, Norwood, MA: Artech House, p. 178.

Long, M. W., 2001, *Radar Reflectivity of Land and Sea*, 3rd ed., Norwood, MA: Artech House. This has always been one of the most comprehensive books. The third edition includes bistatic (physically separate transmit and receive antennas) land and sea clutter.

Nathanson, F. E., 1969, *Radar Design Principles,* New York: McGraw-Hill.

Rheinstein, J., 1968, "Backscatter from Spheres, a Short-Pulse View," *IEEE Trans.,* Vol. AP-l6, Jan., pp. 89–97. (Also 1966, Lincoln Laboratory Technical Report 414, Lexington, MA.) Rheinstein's treatment is exhaustive.

Van Vleck, J. H., F. Block, and M. Hammermesh, 1947, "Theory of Radar Reflection from Wires or Thin Metallic Strips," *Journal of Applied Physics,* Vol. 18, p. 274.

Stimson, George W., 1998, *Introduction To Airborne Radar,* 2nd ed., Raleigh, NC: SciTech Publishing.

5

Waveforms and Signal Processing

5.1 WAVEFORMS

To make the most out of the minute amounts of energy that return to a radar from targets far away depends on intelligent design of radar waveforms. Of prime importance is to extract the signal from the noise (detection), but the waveform must be carefully designed to yield the information (measurements) needed. An infinite number of waveform designs are possible, but the practical designs number in the thousands. However, the essentials can be grasped with only a few illustrations, beginning simply and building on the specifics.

5.2 CHARACTERISTICS OF THE SIMPLE PULSE

A simple radar may be used to measure only range and angle. A simple, short pulse is adequate for this kind of measurement. As the radar becomes more sophisticated, its waveforms will increase in complexity. FAA surveillance and approach control radars may have relatively simple waveforms, whereas those used to support complex weapons systems in the defense department have complex waveforms. Fortunately, the principles governing even the most advanced waveforms are relatively simple and easy to understand, and they can be developed from features of the simple pulse.

Consider a simple pulse composed of a sine wave interrupted by turning a switch on and off. It has a duration τ and an amplitude V, as shown in Figure 5.1. We use a voltage pulse here as it allows for somewhat simpler math. This pulse propagates out to the target, scatters off it, and returns to the radar.

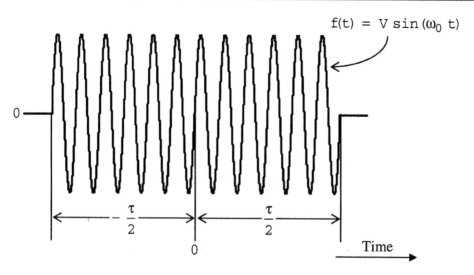

Figure 5.1   A simple pulse—time domain, voltage.

At that time, it has been reduced in energy enormously—by 10 to 30 orders of magnitude (factors of 10) in many cases of interest. Heroic methods are necessary to detect the signal, embedded as it is in the receiver thermal noise.

We can evaluate the simple voltage pulse of Figure 5.1 for its power spectral content (spectrum) by squaring its Fourier transform. The rigorous way to do this is to take the Fourier transform of $f(t) = V \sin(\omega_o t)$ between $-\tau/2$ and $\tau/2$. Stimson (1998, Chapter 17) shows this Fourier transform in all its rigorous detail. This will yield an expression with $\sin(x)/x$ shapes around both $-\omega_o$ and $+\omega_o$. Because we know that, in radar applications, ω_o is on the order of 10^9 Hz, and ωs of interest will be 10% or less of that, we can simply take the Fourier transform of the envelope and remember that, when $\omega = 0$ in the result, it is really at ω_o. Our problem then reduces to taking the Fourier transform of the envelope [Equation (5.1)], which results in Equation (5.2). Using the trig identity for sine [Equation (5.3)], we get the voltage spectrum as given in Equation (5.4). Squaring the voltage spectrum results in the power spectrum [Equation (5.5)].

$$f(t) = V \qquad -\frac{\tau}{2} < V < \frac{\tau}{2} \tag{5.1}$$

$$g(\omega) = \frac{1}{2\pi} \int_{-\tau/2}^{\tau/2} V e^{-j\omega t} dt \tag{5.2}$$

$$\sin(x) = \frac{e^{jx} - e^{-jx}}{2j} \tag{5.3}$$

$$g(\omega) = \frac{V\tau}{2\pi} \frac{\sin\left(\omega \frac{\tau}{2}\right)}{\omega \frac{\tau}{2}} \tag{5.4}$$

$$G(\omega) = \left(\frac{V\tau}{2\pi} \frac{\sin\left(\omega \frac{\tau}{2}\right)}{\omega \frac{\tau}{2}}\right)^2 \tag{5.5}$$

where:

$f(t)$ = time domain representation of the voltage pulse envelope, volts
V = pulse amplitude, volts
τ = pulse duration, seconds
$g(\omega)$ = frequency spectrum of the pulse, volts
ω_o = angular frequency of the pulse, radians/second, $2\pi f_o$, where f_c = carrier frequency, hertz
$G(\omega)$ = frequency spectrum of the pulse, watts

Figure 5.2 is a plot of the power spectrum of the pulse with the root-mean-squared receiver thermal noise power, N, also shown. As the bandwidth of a filter is increased outward from ω_o to ω, signal is added more rapidly than noise for a short distance, after which noise is added more rapidly. It follows that a radar receiver wants to admit only that part of the signal in which gains over noise can be made. Apparently, a filter whose response in spectral content is identical

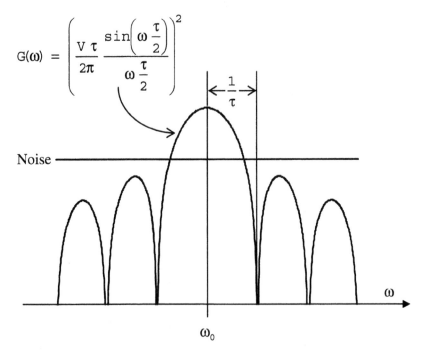

$$G(\omega) = \left(\frac{V\,\tau}{2\pi} \frac{\sin\left(\omega\,\frac{\tau}{2}\right)}{\omega\,\frac{\tau}{2}} \right)^2$$

Figure 5.2   Spectral plot of signal and noise, power, log scale.

to the transmitted pulse will pick up all of the signal if it is not distorted in transit and the noise really has a constant root-mean-square value across the bandwidth (that is, Gaussian white noise). A filter that has this characteristic, and the added one that its phase response is the conjugate of the transmit pulse, is called a *North* [North, 1943] or *matched* filter. A matched filter can be shown to be an optimum filter against the type of noise described. (Mathematical developments of this are frequent. See, for example, Deley, 1964, 1970). The bandwidth of the matched filter is given in Equation (5.6).

$$B = \frac{1}{\tau} \tag{5.6}$$

where:
  B = radar receiver bandwidth, hertz

In practice, a conjugate filter is not necessary and simple third- to fourth-order bandpass filters can be used as matched filters. Notice that their ideal cut-off frequency can be found by solving for the point where the rate of change of integrated power crosses over the rate of change of integrated noise power. Because the integrand is the derivative of the integral in this case, the point is trivial to find, as given in Equation (5.7). The value of receiver thermal noise depends on the expected operating point of the radar. A simple filter such as this has excellent qualities for suppressing noise beyond its passband. As a result, the filter bandwidths of many radars range as given in Equation (5.8) [Barton, 1979, p. 20].

$$\left(\frac{V\tau}{2\pi} \frac{\sin\left(\omega \frac{\tau}{2}\right)}{\omega \frac{\tau}{2}} \right)^2 = N \tag{5.7}$$

$$\frac{0.8}{\tau} < B < \frac{1.2}{\tau} \tag{5.8}$$

where:
  N = radar receiver thermal noise, watts

5.3   RANGE MEASUREMENT

The output of the matched filter is the ratio of peak received target signal power to root-mean-squared noise power (S/N). S/N is the power signal-to-noise ratio and is the criterion for sensitivity applied to all radars. It is easy to see how the output of this matched filter might be hooked into an oscilloscope to display the range of a target to a radar operator. The vertical sweep is connected to the output of the filter and the horizontal sweep to elapsed time, each horizontal sweep beginning at the instant the radar sends out a pulse, as shown in Figure 5.3. This kind of display is often used in radars and is called an "A" scope. The horizontal axis is often calibrated as range, instead of time, so the operator can directly read range off the scope.

Range can also be measured with automatic circuits in the radar receiver. These circuits measure the time delay from when a pulse was transmitted and when the reflected pulse is received (remember the discussion of range in Chapter 1). The range is computed as a function of the measured time delay and speed of light, as given in Equation (5.9). The time delay can also be measured using range gates, or range bins, as shown in Figure 5.4. As shown in this figure, the time bins are usually one pulse width wide. The received target signal will arrive in a specific range gate based on its associated time delay.

$$R = \frac{c\,\Delta t}{2} \tag{5.9}$$

where:

R = radar-to-target range, meters

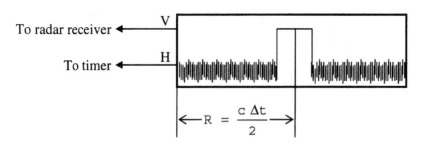

Figure 5.3   A radar "A" scope.

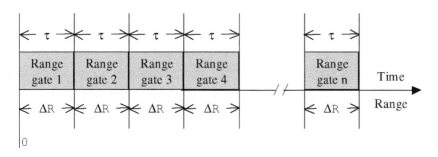

Figure 5.4 Range gates.

Δt = time delay from when a pulse is transmitted and when the reflected pulse is received, seconds

c = speed of light, 3×10^8 meters/second

5.3.1 Resolution

A key question about range measurement is: How far apart in range do targets have to be before they can be resolved? To answer this, it is first necessary to establish some criterion for resolution. When a radar using a matched filter operates on a single pulse in the time domain, operations occur as shown in Figure 5.5. It is apparent that objects separated by more than two pulse widths are absolutely resolvable. In fact, even if noise contaminates the signal, two targets could be resolved if separated at the half-power points (0.707 voltage level) of the pulses, as shown in Figure 5.6. This criterion of resolution is called the "3 dB" (actually –3 dB) or "Rayleigh" criterion. (Lord Rayleigh suggested that the distance from the peak to the first null of the Fraunhoffer diffraction pattern be used as the criterion for resolving one pattern in the presence of another [Jenkins and White, 1976, p. 327]. Radar has simply adopted that approach.) Observe that this time separation is equivalent to the pulse width. Using the relationship between time and range for radar [Equation (5.9)], the resultant range resolution is given in Equation (5.10). A plot of range resolution as a function of pulse width is shown in Figure 5.7. Often, there are practical limits to producing the short pulse width required for fine range resolution, especially for high-power transmitters. If finer range resolution than is practical to achieve with a short pulse width is required, the transmitted carrier can be

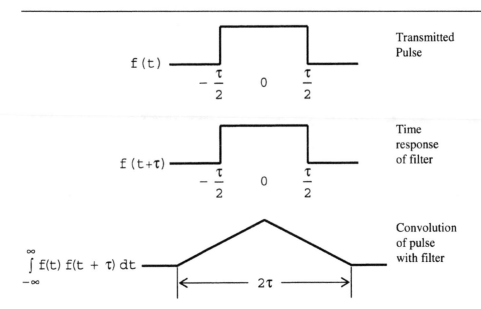

Figure 5.5   Convolution of a simple pulse.

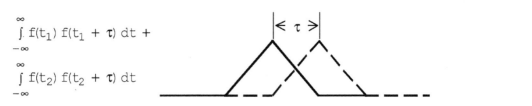

Figure 5.6   Half-power points.

modulated and processed in the receiver, as discussed in the upcoming "Pulse Compression" section.

$$\Delta R = \frac{c\,\tau}{2} \tag{5.10}$$

where:
  ΔR = range resolution, meters
    c = speed of light, 3×10^8 meters/second
    τ = pulse width, seconds

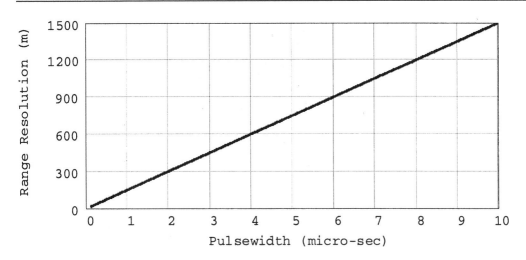

Figure 5.7   Range resolution as a function of pulse width.

5.3.2 Accuracy

Another sensible question to ask about range measurement concerns how accurately it can measure range. An answer is obtained from a great body of work on measurements and probability theory. Accuracy is essentially the statistical measure of how well a measurement can be estimated with its resolution. For example, a meter stick that is marked in centimeters has a resolution of one centimeter. We can estimate better than one centimeter, but it is just that—an estimate—and it is a function of other factors. The usual equation for the RMS range accuracy potential is given in Equation (5.11). All the accuracy expressions derived in this section are within a few percent of the maximum likelihood estimates of statistics, which are optimum. (Manasse, 1960, gives the exact equations; Barton, 1979, pp. 38–43, explains the differences.) When derived in more sophisticated ways and optimized by various methods, this result changes by small amounts.

$$\delta_R = \frac{\Delta R}{\sqrt{2\dfrac{S}{N}}} = \frac{c\tau}{2\sqrt{2\dfrac{S}{N}}} = \frac{c}{2B\sqrt{2\dfrac{S}{N}}} \tag{5.11}$$

where:

δ_R = RMS range accuracy, meters

S/N = signal-to-noise ratio (power), no units

5.3.3 Ambiguity

Since the time delay is measured with respect to the time a pulse was transmitted, only time delays less than the pulse repetition interval (PRI) are unambiguous. Time delays greater than the *PRI* are incorrectly interpreted, as shown in Figure 5.8. The distance to which a pulse can make a round trip before the next pulse is sent is called the "unambiguous" range or "first time around" range. The unambiguous range is given in Equation (5.12). A plot of unambiguous range as a function of the pulse repetition frequency (PRF) is shown in Figure 5.9.

$$R_u = \frac{c \, PRI}{2} = \frac{c}{2 \, PRF} \tag{5.12}$$

where:

R_u = unambiguous range, meters

c = speed of light, 3×10^8 meters/second

PRI = pulse repetition interval, seconds

PRF = pulse repetition frequency, hertz

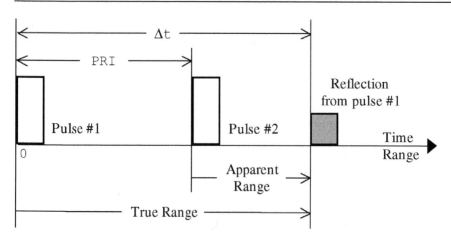

Figure 5.8 Unambiguous range.

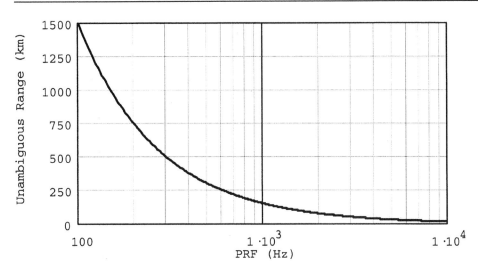

Figure 5.9 Unambiguous range as a function of pulse repetition frequency.

5.3.4 Dead Zone

Normally, the radar receiver is shut off or isolated while the transmitter is on. No targets can be detected during this interval. In the meantime, the leading edge of the pulse has traveled out to $c\tau$ from the transmitter. The radius of what is called the "dead zone" is $c\tau/2$. The zone is not really dead, because special arrangements can be made so that some targets can be detected with the trailing portions of the transmitted pulses.

5.3.5 Eclipsing

Eclipsing occurs when a reflected pulse is received while the transmitter is on (and the receiver is not connected to the antenna). Eclipsing is shown in Figure 5.10. Eclipsing results from targets with a time delay approximately equal to the PRI. Eclipsing results in a loss of power in the received pulse, which affects detection performance. The entire received pulse can be eclipsed (full eclipsing), or only part of the received pulse can be eclipsed (partial eclipsing). The lower the PRI, higher the PRF, the more likely eclipsing will impact the radar. The effects of eclipsing can be minimized through the choice of PRF/PRI and/ or radar detection performance.

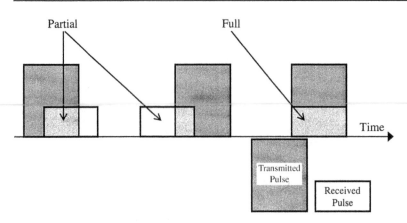

Figure 5.10 Eclipsing.

5.4 DOPPLER MEASUREMENTS

Still another question of interest about the single simple pulse is its Doppler measuring capabilities. The Doppler frequency can be derived from first principles, assuming that the speed of the object is much less than the speed of light. An object is moving toward a radar at a certain range rate—time rate of change of range. The range rate, or speed, is the magnitude of the projection of the object's velocity vector on the object-to-radar range vector. The radar is propagating electromagnetic energy at the object at its carrier frequency. An object moving toward the radar will intercept more cycles of electromagnetic energy in 1 sec. For example, if the object is moving at 300 m/sec, and the wavelength is 0.3 m, 90 additional cycles will be intercepted in 1 sec, and the frequency of the radar will seem at the object to be the carrier plus 90. The object now reradiates the carrier plus 90 cycles back toward the radar. But in so doing, because of the object's range rate, each successive cycle is reradiated at a point slightly closer to the radar than the previous one, and the wavelength is again compressed so that 90 additional cycles occur in 1 sec. These two added increments of frequency are called the *Doppler shift*, as given in Equation (5.13). The Doppler shift is measured in the radar receiver and used to compute the range rate, as given in Equation (5.14).

$$f_d = \frac{2R_{dot}}{\lambda} \tag{5.13}$$

$$R_{dot} = \frac{f_d \lambda}{2} \tag{5.14}$$

where:

f_d = Doppler shift, hertz

R_{dot} = object-to-radar range rate, meters/second

λ = wavelength, meters

5.4.1 Resolution

To find the Doppler resolution of the simple pulse, it is necessary to return to the frequency domain (see the sketches in Figure 5.11). In keeping with the criterion established for range resolution, it can be stated that two objects are resolved in Doppler when their peaks are separated by one over the pulse width.

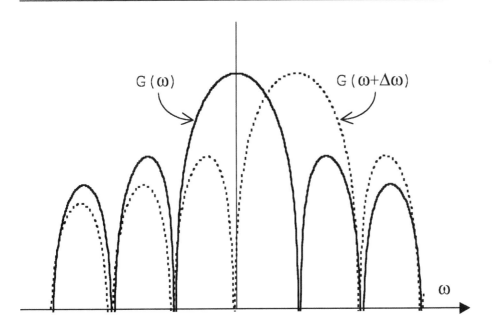

Figure 5.11 Doppler resolution, simple pulse.

This is approximately the 3-dB point. Thus, the Doppler resolution is given in Equation (5.15). This equates to a range rate resolution as given in Equation (5.16). When targets of unequal RCS are of interest, much more stringent bases for resolution are adopted, and tapering is invoked to decrease the Doppler sidelobes, with the idea of resolving targets that are many thousands of times different in amplitude. (This is exactly as it was in Chapter 2 to suppress the angle sidelobes of antennas.)

$$\Delta f_d = \frac{1}{\tau} \tag{5.15}$$

$$\Delta R_{dot} = \frac{\lambda}{2\tau} \tag{5.16}$$

where:
Δf_d = Doppler resolution, hertz
ΔR_{dot} = range rate resolution, meters/second

In the previous range measurement section, we found that the range resolution was also a function of the pulse width. A finer range resolution requires a short pulse width, while a finer range rate resolution requires a long pulse width. Clearly, the radar designer has a trade-off to make when dealing with a simple pulse for range and range rate resolution.

5.4.2 Accuracy

The accuracy limits for Doppler frequency and range rate can be found by applying the same methods as were used for finding the accuracy limits for range, as given in Equations (5.17) and (5.18).

$$\delta f_d = \frac{\Delta f_d}{\sqrt{2\frac{S}{N}}} = \frac{1}{\tau\sqrt{2\frac{S}{N}}} \tag{5.17}$$

$$\delta R_{dot} = \frac{\Delta R_{dot}}{\sqrt{2\frac{S}{N}}} = \frac{\lambda}{2\tau\sqrt{2\frac{S}{N}}} \tag{5.18}$$

where:

δf_d = RMS Doppler accuracy, hertz

δR_{dot} = RMS range rate accuracy, meters/second

5.5 ANGLE MEASUREMENTS

Angle measurements are performed using the radar antenna. Angle resolvers tell the radar system where a mechanically scanned antenna is pointing. The beam steering computer tells the radar system where an electronically steered antenna beam is pointing. In keeping with the criterion established for range resolution, it can be stated that two objects are resolved in angle when their peaks are separated by one antenna 3-dB beamwidth. Thus, the angular resolution is given in Equation (5.19).

$$\Delta\theta = \theta_{3dB} \tag{5.19}$$

where:

$\Delta\theta$ = angle resolution, degrees or radians

θ_{3dB} = antenna half-power (–3dB) beamwidth, degrees or radians

Good monopulse trackers can easily get accuracies of one-tenth of the radar beamwidth. In fact, they can come close to achieving the theoretical angular accuracy capability of a radar. This theoretical performance can be derived from first principles, as given in Equation (5.20).

$$\delta_\theta = \frac{\theta_{3dB}}{\sqrt{2\frac{S}{N}}} \tag{5.20}$$

where:

δ_θ = angle measurement accuracy, radians or degrees

For S/N = 17 dB, beam splitting of ten to one is accomplished. The better tracking radars (FPS-16 range measurement systems) achieve 0.1-mrad accuracies at C-band with 4.6-m reflector antennas.

5.6 PULSE COMPRESSION

For a simple pulse, the pulse's duration determines its range resolution and accuracy, and the reciprocal of its duration determines its Doppler resolution and accuracy. Therefore, it is natural to conclude that high resolution in range and Doppler cannot be obtained simultaneously. Such is not the case. By a set of techniques called *pulse compression* [Stimson, 1998, Chapter 13; Skolnik, 2001, Section 6.5], a single pulse can provide both good range and Doppler resolution [Farnett, Howard, and Stevens, 1970]. We shall present one way of accomplishing pulse compression in detail and allude to several others.

5.6.1 FM Chirp

The pulse compression technique to be explained is called *FM chirp,* with *FM* for frequency modulation and *chirp* for the fact that a ramp waveform that changes in frequency is used; if it were in the audible range, it would make a chirping sound.

FM chirp is conceptually simple, and the range resolution it achieves is easy to understand. We are looking for a pulse with both a wide bandwidth and a substantial duration. One such pulse is shown in Figure 5.12. When this pulse is transmitted, a replica of it could be stored and mixed with a pulse returning from a target. The result, if the replica ramp were extended at both ends beyond the duration of the return pulse, would appear as shown in Figure 5.13.

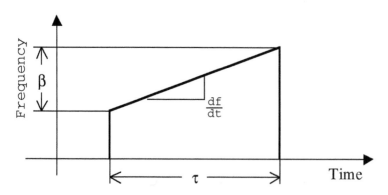

Figure 5.12 FM chirp pulse compression.

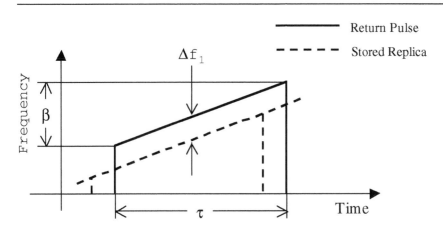

Figure 5.13 Difference frequency generation.

The return pulse of one target is shown, and one product of the mixture would be the difference frequency between the two ramps, Δf_1. Because Δf endures for the pulse width, it has the characteristics shown in Figure 5.14. Another target separated by Δf_2 would be resolved, providing it is separated from Δf_1 by $1/\tau$ cycles. The bandwidth of the pulse is thus divisible into resolution elements of $1/\tau$

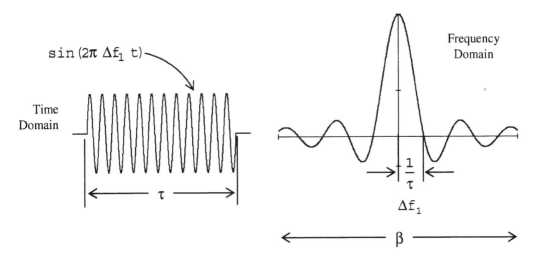

Figure 5.14 Spectrum of a difference frequency.

cycles. The number of resolution cells is the bandwidth of the pulse divided by the number of resolution elements, or the time-bandwidth product. The time-bandwidth product is given in Equation (5.21). The time-bandwidth product is often called the *pulse compression ratio (PCR)*, the ratio of the transmitted pulse width to the compressed pulse width.

$$\frac{\beta}{1/\tau} = \beta\tau \qquad (5.21)$$

where:

  β = frequency excursion of the FM pulse, hertz

An alternative form of pulse compression processing is to use a filter whose transit time is proportional to frequency. Such a filter and its ability to compress the pulse width of a pulse with a frequency modulated carrier are shown in Figure 5.15. The filter provides a time delay that decreases linearly with frequency at the same rate that the carrier frequency of the pulse increases. Thus, the higher frequencies associated with the trailing portions of the pulse take less time to pass through the filter than the lower frequencies associated with the leading portion. Successive portions of the pulse will essentially stack up at the output of the filter.

Consequentially, when the pulse emerges from the filter its amplitude is greater and its width is less than when it entered. The pulse still has the same

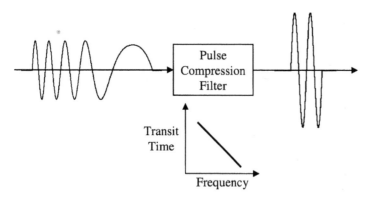

Figure 5.15 Pulse compression processing.

energy, but the pulse width has been compressed. Such a pulse compression filter can be implemented with analog devices, such as an acoustical delay line, or digitally.

The amplitude of the simple pulse is increased by the time-bandwidth product. To ensure conservation of energy of the simple pulse, the duration of the simple pulse is reduced by the time-bandwidth product. The previous simple pulse range resolution is now increased (made finer) by the time-bandwidth product, as given by Equation (5.22). Equation (5.22) shows that the range resolution of a pulse can be independent of its duration. Duration, of course, is selected for other reasons (for example, to increase average power or improve Doppler resolution). A plot of range resolution as a function of pulse width is shown in Figure 5.16.

$$\Delta R = \frac{c}{2\beta} \tag{5.22}$$

5.6.2   Features of FM Chirp

FM chirp provides another example of what is called (erroneously) "signal processing gain," the idea being that, by signal processing, a gain in range resolu-

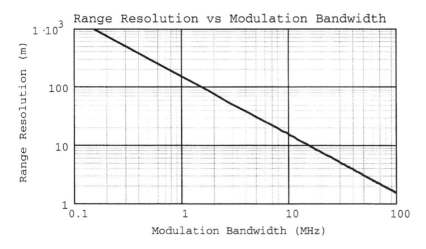

Figure 5.16   Range resolution as a function of pulse bandwidth.

tion of a factor of the time-bandwidth product over the simple pulse has been obtained. In actuality, the frequency modulation on the pulse is used to compress the pulse duration, thus resulting in a gain in range resolution.

FM chirp has an unfortunate characteristic: If the target has an unknown Doppler shift (known Doppler shifts can be subtracted out), that Doppler shift contributes to Δf and causes an error in the measurement of range. Objects separated in range by ΔR but separated in Doppler by Δf might be superimposed in the output of the pulse compression filter, as given in Equation (5.23).

$$\Delta f_{total} = \Delta f_{range} + \Delta f_{Doppler} \tag{5.23}$$

where:

Δ_f = total frequency difference, hertz
Δf_{range} = frequency difference due to the range of the target, hertz
$\Delta f_{Doppler}$ = frequency difference due to the Doppler shift of the target, hertz

These kinds of errors are readily quantified. The target range is determined by the magnitude of Δf compared to the reference ramp. The error in this range caused by target Doppler is just the apparent displacement in range that the Doppler causes, which, in terms of range resolution bins, is given by Equation (5.24). The conversion of range resolution bins to range is given in Equation (5.25). Thus, Equation (5.25) is the error in range due to an unknown target Doppler shift.

$$N_{bins} = \Delta f_{Doppler} \, \tau \tag{5.24}$$

$$\varepsilon_R = \Delta f_{Doppler} \, \tau \, \frac{c}{2\beta} \tag{5.25}$$

where:

N_{bins} = displacement in the number of range bins due to target Doppler, no units
τ = transmitted pulse width, seconds
ε_R = error in range due to unknown Doppler shift, meters
c = speed of light, 3×10^8 meters/second

For some radar waveforms, unknown target Doppler shifts are not a problem; for others, it is severe. Consider a ballistic missile measurement radar that has a

20-μsec pulse and a 10-MHz FM chirp at L-band. This results in a range resolution of 15 m. The uncertainty in target Doppler for ICBM trajectories might be 5 kHz (the absolute f_d about 50 kHz). The error in range due to the Doppler shift is given in Equation (5.26).

$$\varepsilon_R = (5000)(20 \times 10^{-6})\frac{3 \times 10^8}{2(10 \times 10^6)} = 1.5 \text{ m} \tag{5.26}$$

Another measurements radar might have a waveform with a 50-μsec pulse and a 1-MHz FM chirp at UHF. This results in a range resolution of 150 m. The uncertainty in target Doppler for ICBM trajectories might be 20 kHz. The error in range due to the Doppler shift is given in Equation (5.27).

$$\varepsilon_R = (20 \times 10^3)(50 \times 10^{-6})\frac{3 \times 10^8}{2(1 \times 10^6)} = 150 \text{ m} \tag{5.27}$$

Although the gross error is less than the range accuracy error we might expect for this system, the error in placement of objects with respect to other objects could be serious, and careful data processing is required if it is likely that objects of widely varying range rates will be within the same pulse. The Doppler crosstalk in FM chirp can be eliminated by alternating up ramps with otherwise identical down ramps, pulse-to-pulse. The forward displacement in the up ramp becomes a backward displacement in the down ramp, the actual position being the arithmetic mean. Depending on the need, the added complexity and expense in electronics and software may not be worthwhile.

5.6.3 Alternate Methods

Wide varieties of additional schemes exist for synthesizing wideband pulses that are also of long duration. One is to generate wideband CW noise, simultaneously storing it for use as a reference on receive. Some isolation from the transmitter is achieved, because the waveform being transmitted is decorrelated with that being received. Arbitrary lengths of the return signal are matched against the reference waveform. The range resolution achieved is proportional to the noise bandwidth, and the processing bandwidth is equal to the reciprocal of the pulse duration. While the principal response of this waveform may be optimum, range sidelobe characteristics will be volatile because of the random structure of the signal. The approach is similar to spread spectrum transmission in communications. Its conceptual appeal still outstrips practical results, although recent advancements in processing capability have resulted in renewed interest.

Another technique is to divide a pulse into several sub-pulses of different center frequencies, each with a bandwidth equal to the reciprocal of the duration of the whole pulse [Stimson, 1998, p. 164]. If the sub-pulses are transmitted in sequence with their frequency steps, the signal is like a stepped FM chirp. If they are transmitted randomly, they are similar to a pseudorandom noise signal. The sub-pulses can be separated in time to become a series of pulses, each at a different center frequency, having the effect of increasing time in the time-bandwidth product. If gaps are left between the frequencies of the sub-pulses, ambiguities appear, complicating processor design.

A further approach is to divide a long pulse into coded segments by shifting the phase, often called *phase modulation*. Biphase (180° phase shifts) is the usual, but by no means the only, amount. When a segment is autocorrelated with a segment 180° out of phase, the two cancel; otherwise, they give a response. By carefully selecting segment phases, these responses can be kept to a height of unity except for the maximum response that is equivalent to the number of segments in the code. In other words, the range sidelobes have unit height. Such a code is called a *Barker code*. The longest Barker code so far discovered is 13—not really useful for pulse compression in most modern radars [Skolnik, 2001, pp. 350–351, 361–362; Stimson, 1998, p. 172]. Practical phase codes with larger numbers of segments use pseudorandom phase selection techniques leavened with experimentation. Polyphase (<180° phase shifts, e.g., 90°, 45°, and so on) codes have come into use with modern radar systems. While not providing the uniform range sidelobes of a Barker code, they can provide a much finer range resolution by dividing a long pulse into many coded segments. The range resolution achieved by using phase modulation, and the resultant pulse compression processing is also given in Equation (5.22). The bandwidth of a phase modulated pulse is given in Equation (5.28).

$$\beta = \frac{N\phi}{\tau} \qquad\qquad (5.28)$$

where:

   β = bandwidth of a phase modulated pulse, hertz

   $N\phi$ = number of phase coded segments, no units

   τ = transmitted pulse width, seconds

5.6.4   Impact of Pulse Compression on S/N

As previously stated, the signal amplitude is increased by the time-bandwidth product because of pulse compression processing. Thus, at first glance, one would think the S/N would increase. However, the receiver bandwidth must be wide enough to pass the modulation bandwidth of the pulse. Thus, because the receiver thermal noise is a function of the receiver bandwidth, the receiver thermal noise is also increased. The S/N after pulse compression processing is the same as the input S/N [Goj, 1993]. Conservation of energy holds true. This will be shown in an exercise at the end of this chapter.

5.6.5   Range and Doppler Sidelobes

Single pulses with high time-bandwidth products (the term applied to a high-resolution, long-duration waveform) are very powerful signal-processing tools. Nevertheless, they have certain deficiencies that make them less than ideal when used in a cluttered environment such as one would expect to encounter in a full-scale test on the range or in an actual engagement. These deficiencies arise because the single-pulse waveform has sidelobes in both the range (time) and Doppler dimensions. Recall that in the range direction, along the reference Doppler track, the compressed pulse has the form of $(\sin x/x)^2$, which has monotonically decreasing sidelobes beginning at about 13.4 dB down from the peak response, as shown in Figure 5.17. If the target of interest has a small S/N and a target, say 30 or 40 dB larger, is in a nearby range gate, the spillover of the range sidelobes from the stronger target into the range gate of interest masks the small S/N target. Because the Doppler response is also of the $(\sin x/x)^2$ form, the sidelobes in Doppler are equally troublesome when a large target exists in a nearby Doppler filter.

   To counteract these effects, clever "weighting functions" are often applied to the simple pulse, as they are to the antenna illumination functions discussed in Chapter 2. If the transmitted pulse is $\cos(xt)$ instead of $V(t)$, as shown in

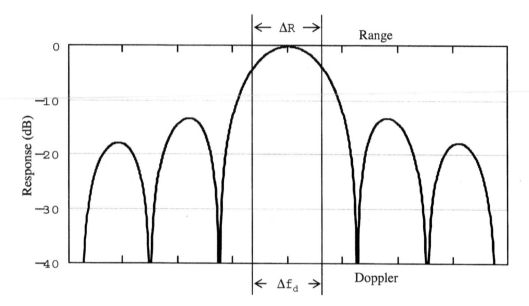

Figure 5.17 Power response of sidelobes.

Figure 5.18, the first sidelobe of this function is down 23 dB from the peak response. The main lobe, however, is 37% wider than for the simple pulse. As suggested in Chapter 2, one can use different weighting functions to achieve certain sidelobe levels. Some of these weighting functions have very low sidelobes, achieved at great penalty in resolution (due to the widening of the main lobe). In experimenting, one may run across weighting functions that already have famous names. For example, $C_1 + C_2 \cos(xt)$ is the Hamming weighting; its

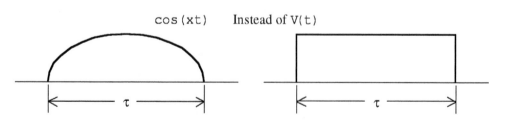

Figure 5.18 A cosine weighted pulse.

first sidelobe is >43 dB down [Temes, 1962]. Because the Fourier transform of the Gaussian distribution is another Gaussian distribution, that waveform has no time sidelobes whatsoever and, of course, is also physically unrealizable.

5.7 PULSE BURST WAVEFORMS

For the reasons previously discussed, and other reasons involving radar operations themselves, multiple-pulse and pulse burst waveforms have been invented. These can be tailored to the precise needs of the radar using them. Pulse-Doppler radars effectively achieve pulse bursts by simply breaking up their continuous streams of pulses into intervals for processing. Instrumentation radars at this country's various missile and aircraft test ranges use a variety of pulse bursts.

In the design of a multiple-pulse waveform, exact knowledge of the mission is required so that unambiguous Doppler and range intervals may be selected and appropriate range and Doppler resolutions may be obtained. To see how this is done, it is necessary to view again the waveform in both the time and frequency domains. To extract Doppler information from a pulse burst waveform, each pulse must be coherent with respect to the next. Consider a pulse train in time and its Fourier transform (spectrum), as shown in Figure 5.19. Notice how the spectrum of a coherent pulse burst waveform is a sequence of spectral lines. We will see how measurement characteristics relate to pulse burst waveform characteristics, as shown in Table 5.1.

Table 5.1 Measurement and Waveform Characteristics

Range resolution	τ
Doppler resolution	T
Range ambiguity	PRI
Doppler ambiguity	$PRF = 1/PRI$

A pulse burst waveform permits arbitrarily good range and Doppler resolution; that is, the pulse can be as short as necessary (or each can be given pulse compression) while the pulse burst duration (Doppler processing time) can

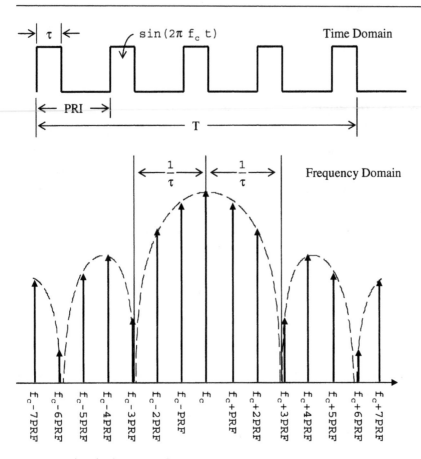

Figure 5.19 A pulse burst waveform and its spectrum.

be as long as necessary. These qualities are essential for pulse-Doppler, MTI, radar mapping, and ballistic missile defense radars, but they are not an unvarnished blessing. With them comes the problems of range and Doppler ambiguities, so called because the range or the Doppler or both may be uncertain to some degree, and because of blind ranges and speeds, which will not be seen by the radar.

In Figure 5.19, the PRI is long enough to make a round trip to the unambiguous range R_u [Equation (5.12)], which is fine for targets within that range. If, however, there are targets of interest farther out, inside two, three, or four times R_u, pulses will return from those targets and, once they start appearing, will show up in every PRI. All targets, regardless of range, will appear within all PRIs. Targets will be ambiguous as to which PRI they are in, although their position within the PRI itself will be known accurately.

The Doppler shift associated with a moving target is measured using a bank of narrow bandpass filters, as shown in Figure 5.20. The Doppler filter containing the target Doppler shift will have the strongest output signal. Narrowband filters achieve their selectivity by integrating the input signal over a period of time, as given in Equation (5.29). Thus, the longer the integration time, the smaller the Doppler filter bandwidth. Range rate resolution is a function of the Doppler filter bandwidth. Two Doppler shifts can be resolved if they are separated in frequency by one Doppler filter bandwidth. Converting from the frequency domain to the range rate domain provides the range rate resolution, as given in Equation (5.30). The range rate resolution per unit Doppler filter bandwidth as a function of radar carrier frequency is shown in Figure 5.21.

$$\Delta f_d = \frac{1}{T} \tag{5.29}$$

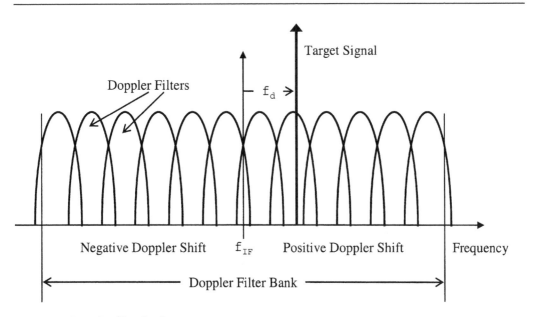

Figure 5.20 Doppler filter bank.

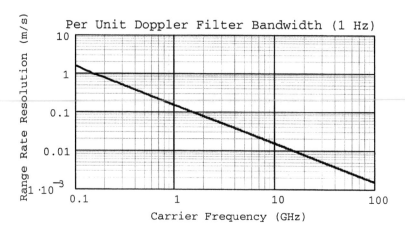

Figure 5.21 Range rate resolution as a function of carrier frequency.

$$\Delta R_{dot} = \frac{\Delta f_d \ \lambda}{2} \qquad (5.30)$$

where:

Δf_d = Doppler filter bandwidth, hertz
T = pulse burst time, seconds
ΔR_{dot} = range rate resolution, meters/second
λ = wavelength, meters

Frequency domain ambiguities exist for a pulse burst waveform. The spectrum of the received signal from a moving target is a Doppler shifted version of the transmitted spectrum, as shown in Figure 5.22. The Doppler ambiguity is a function of whether the radar can detect only positive Doppler shifts or both positive and negative Doppler shifts. For targets with only positive Doppler shifts (closing on the radar), all is well and good for Doppler shifts less than the PRF. If, however, there are Doppler shifts greater than the PRF, they will appear between the carrier and the spectral line associated with the first PRF in a position denoted by the excess of their Doppler shift beyond a multiple of the PRF. Target Doppler shifts will be ambiguous with respect to the carrier, although their position between the carrier and the first PRF will be known accurately. Thus, for targets with only positive Doppler shifts, the unambiguous Doppler

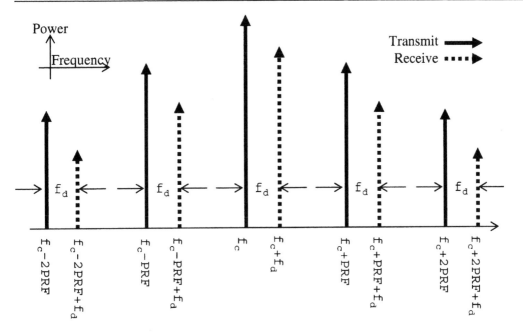

Figure 5.22 Transmitted and received spectra.

shift is the PRF, as given in Equation (5.31). Converting from the frequency domain to the range rate domain provides the unambiguous range rate, as given in Equation (5.32). Doppler ambiguities are often referred to as "foldover" because Doppler shifts outside a valid range fold over into the valid range.

$$f_{du+} = PRF \tag{5.31}$$

$$R_{dotu+} = \frac{f_{du+}\,\lambda}{2} = \frac{PRF\,\lambda}{2} \tag{5.32}$$

where:

f_{du+} = unambiguous Doppler shift—positive Doppler shifts only, hertz
R_{dotu+} = unambiguous range rate—positive Doppler shifts only, meters/second

The much more common radar capability is to be able to detect targets with both positive and negative Doppler shifts (closing on and opening from the radar). For targets with both positive and negative Doppler shifts, all is well and good for Doppler shifts within ±PRF/2. If, however, there are Doppler shifts outside ±PRF/2, they will appear between −PRF/2 and +PRF/2 in a position denoted by the excess of their Doppler shift beyond a multiple of the PRF, as shown in Figure 5.23. Target Doppler shifts will be ambiguous with respect to the reference frequency (more on this later), although their position between −PRF/2 and +PRF/2 will be known accurately. Thus, for targets with positive and negative Doppler shifts the unambiguous Doppler shift is the ±PRF/2, as given in Equation (5.33). Converting from the frequency domain to the range rate domain provides the unambiguous range rate, as given in Equation (5.34). The unambiguous range rate per unit PRF as a function of radar carrier frequency is shown in Figure 5.24.

$$f_{du} = \pm \frac{PRF\,\lambda}{2} \tag{5.33}$$

$$R_{dotu} = \frac{f_{du}\,\lambda}{2} = \pm \frac{PRF\,\lambda}{4} \tag{5.34}$$

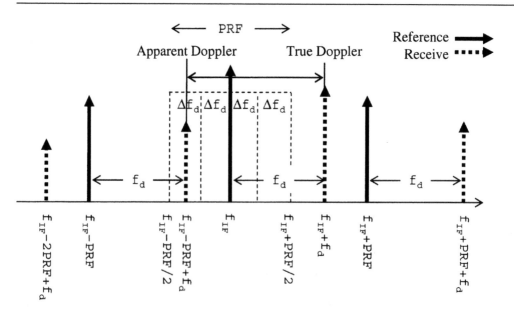

Figure 5.23 Ambiguous Doppler shift.

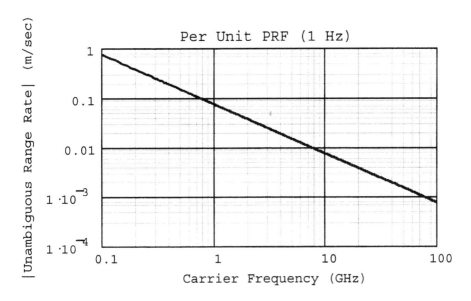

Figure 5.24 Unambiguous range rate as a function of carrier frequency.

where:

f_{du} = unambiguous Doppler shift—positive and negative Doppler shifts, hertz

R_{dotu} = unambiguous range rate—positive and negative Doppler shifts, meters/second

An example will make these ambiguities more concrete. Let us say that we wish to keep track of Mach 8 missiles coming toward a ship (positive Doppler shifts only) from a 500-km range with an L-band ($\lambda \approx 0.3$ m) radar. The perfect waveform would have its range rate ambiguity beyond Mach 8 (2640 m/sec) and its range ambiguity beyond 500 km. Let us calculate those. Because we are interested only in positive Doppler shifts, the PRF necessary to provide the range rate ambiguity is computed using Equation (5.32) (solved in terms of PRF), with the result given in Equation (5.35). The PRF necessary to provide the range ambiguity is computed using Equation (5.12) (solved in terms of PRF), with the result given in Equation (5.36). Clearly, the two requirements are incompatible. The designer must accept either range or Doppler ambiguities.

$$PRF = \frac{2R_{dotu}}{\lambda} = \frac{2(2640)}{0.3} = 17.6 \text{ kHz} \tag{5.35}$$

$$PRF = \frac{c}{2R_u} = \frac{3 \times 10^8}{2(500 \times 10^3)} = 300 \text{ Hz} \tag{5.36}$$

A frequently used way around range and Doppler ambiguities is to use multiple PRFs [Skolnik, 2001, pp. 175–178; Stimson, 1998, p. 155, 286]. The result will be that ambiguous ranges will change position, and the actual range will remain fixed. A coincidence filter can sort out the ambiguities, providing there are not too many. If ambiguities still exist, a third PRF can be generated. The procedure is shown in Figure 5.25. Doppler ambiguities can be resolved in a like manner.

A radar that can use a long train of multiple pulses such as that previously described must be able to preserve phase throughout the generation, transmission, and reception of the waveform so that all the carrier and modulation phases will add up correctly. This quality, which requires the radar to operate with phase-locked loops and a stable master clock, is *coherence*. It is mandatory for modern radars.

The relationships between radar waveform characteristics and measurements discussed in the previous sections are shown in Table 5.2.

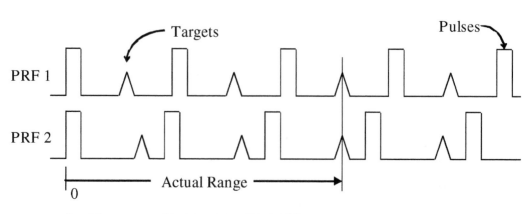

Figure 5.25 Resolving range ambiguities with multiple PRFs.

Table 5.2   Waveform and Measurement Relationships

Parameter	Simple Pulse	Compressed Pulse	Pulse Burst
Range resolution—ΔR	$\dfrac{c\,\tau}{2}$	$\dfrac{c}{2\beta}$	$\dfrac{c\,\tau}{2}$
Range accuracy—δR	$\dfrac{c\,\tau}{2\sqrt{2\dfrac{S}{N}}}$	$\dfrac{c}{2\beta\sqrt{2\dfrac{S}{N}}}$	$\dfrac{c\,\tau}{2\sqrt{2\dfrac{S}{N}}}$
Range ambiguity—R_u	$\dfrac{c\,PRF}{2} = \dfrac{c}{2\,PRF}$		
Range rate resolution—ΔR_{dot}		$\dfrac{\lambda}{2\tau}$	$\dfrac{\Delta f_d\,\lambda}{2}$
Range rate accuracy—δR_{dot}		$\dfrac{\lambda}{2\tau\sqrt{2\dfrac{S}{N}}}$	$\dfrac{\Delta f_d\,\lambda}{2\sqrt{2\dfrac{S}{N}}}$
Range rate ambiguity—R_{dotu+} (only positive Doppler shifts)		None	$\dfrac{PRF\,\lambda}{2}$
Range rate ambiguity—R_{dotu+} (only positive and negative Doppler shifts)		None	$\pm\dfrac{PRF\,\lambda}{4}$
Angle resolution–$\Delta\theta$	$\theta_{3\,dB}$		
Angle accuracy—$\delta\theta$	$\dfrac{\theta_{3\,dB}}{\sqrt{2\dfrac{S}{N}}}$		

5.8   AMBIGUITY FUNCTIONS

When the range axis and the Doppler axis are combined, they form a plane out of which the responses of various waveforms rise. The resulting three-dimensional surface is called the *Woodward ambiguity function,* after Woodward (1955), whose slender volume is an elegant and classical work. The ambiguity function of a simple pulse is shown in Figure 5.26. A whole sector of radar engineering, involving many people working over several decades, has been devoted to manipulating the ambiguity function to obtain an infinitude of desired results. Rihaczek (1977) contains ambiguity functions for many radar waveforms.

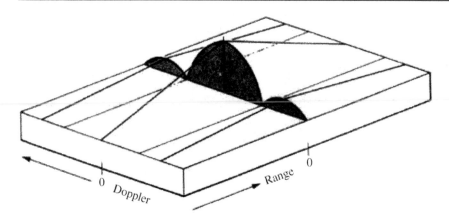

Figure 5.26 Ambiguity function for a simple pulse.

5.9 SIGNAL PROCESSING

Waveforms are the tools that give the radar the means for accomplishing its mission. Signal processing is the sequence of operations performed on those waveforms to extract and display or store essential information. Although powerful digital computers have become a major force in signal processing in modern sophisticated radars, a substantial portion remains analog—for both practical and historical reasons. Due to the complexity of the subject and the proliferation of clever ideas, separating concept from technique is often difficult. The approach in this section is to explain signal processing in a basic radar system and follow with a discussion of several more sophisticated processes.

5.9.1 Basic Receiver Chain

All radar operations are slaved to a master oscillator (clock). When a radar pulse is transmitted, counting begins. When the pulse clears the antenna, it is routed to the receiver chain. Most radars use a modified form of a superheterodyne receiver (similar to, say, a table or car AM/FM clock radio), a block diagram of which is shown in Figure 5.27.

The first radio frequency (RF) amplifier, in any system requiring high sensitivity, is a low-noise amplifier inserted as close as possible to the receiving an-

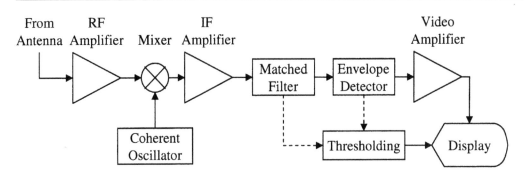

Figure 5.27 Basic radar receiver chain.

tenna to minimize the noise that enters the system prior to first amplification. The mixer translates the carrier frequency down to an intermediate frequency (IF) to facilitate handling while preserving phase and other signal characteristics.

At IF, matched filters, which are optimum for selecting signal against noise, are inserted. All operations on the signal that require preservation of IF phase inside a pulse or from pulse to pulse must be conducted here. Examples of operations that require the preservation of phase are coherent integration, coherent sidelobe cancellation, some forms of pulse compression, and Doppler processing.

At the envelope detector, the signal is further filtered. IF phase information is lost, leaving only the modulation envelope (the video). It is amplified in the video amplifier for further processing of several kinds. Often, one or more of the feedback loops that permit the receiver to operate over very large dynamic ranges is applied here. With R^4 signal strength dependency, and large fluctuations in RCS, signals arriving in the amplifier chain may vary in amplitude by 80 dB or more.

Different aspects of this problem of large signal dynamic range are attacked by three popular techniques: automatic gain control (AGC), sensitivity time control (STC), and constant false alarm rate (CFAR) processing.

AGC is a lot like automatic volume control in a radio. Part of the signal voltage in an amplifier stage is tapped off and fed back to preceding stages as a negative bias, making the chain self-damping at a tolerable signal level. Because

AGC involves a feedback loop operating continuously and adaptively, real signal levels are lost, with obvious effects on a radar designed for precision measurements. STC is synchronized with the radar's master clock, applying exponential attenuation (or gain) to the receiver chain based on time elapsed since a pulse was transmitted. In this way, return signals from nearby targets are heavily attenuated (or have minimal gain), while those from far away have no attenuation (or maximum gain). Fourth- and third-power exponential attenuation (or gain) functions are common. CFAR is a more complex technique. It involves sampling and averaging the noise level in regions near the target and adaptively setting detection thresholds at that level. CFAR relieves the system from processing excess false alarms but may also prevent the detection of signals. Because CFAR keeps displays clear in a jamming environment, it is sometimes falsely credited with anti-jam capability.

The video amplifier also drives the radar displays. Cathode ray tubes (CRTs), which are similar to home television screens, are used in abundance by radars. Even today, the deflection plates of many CRTs are coupled to the radar's master clock, or the antenna pedestal (or the phased array's beam steering computer), and the incoming signal. Often, however, the graphics for modern displays are synthesized from the real data by the radar's computer.

The plan position indicator (PPI) is one of the most common displays. Used with surveillance radars, it gives a plan view of the radar coverage: radar in the center of the scope, several concentric rings marking range, with azimuth read directly from a compass rose around the edge of the CRT. The persistence of the screen's phosphors keeps targets visible between scans, but the operator may still use grease pencil or a light pen for tracking.

The so-called A-scope shows range as the abscissa of the display with signal amplitude as the ordinate. In like manner, the B-scope displays range as a function of angle. An E-scope displays range as a function of elevation angle. "Official" displays are numerous [Skolnik, 2001, Section 11.5; Stimson, 1998, p. 21], but many are not widely used or have fallen into disuse.

With modern computer graphics, all of the above displays could be made available on a single screen, either in time multiplex or with a split screen presentation. In fact, the PPI could be combined with the E-scope in a three-dimensional (3D) perspective display. More advanced concepts, such as holographic displays that provide what seems to be a true 3-D presence, are also receiving attention.

5.9.2 Signal Processing Topics

Many of the multitudes of signal processing capabilities available in modern radar are discussed conceptually elsewhere in this book. In this section, we give some detailed attention to the mechanics of implementing multiple pulse integration processing and Doppler processing, including digital pulse Doppler processing.

5.9.2.1 Integration. Integration can take place at IF before detection or after the envelope detector. Integration at IF is called *predetection* integration. After the envelope detector, it is called *postdetection* or *envelope* integration. Predetection integration can be done with a range gate, a narrowband filter, and a thresholding device, as shown in Figure 5.28. There may be many such channels in parallel, sampling different range gates. Range gates are merely switches that open to receive the return from a target (if any) at a particular range. Their purpose is to limit the signal passed to the filter to that associated with a given range. The thresholding device declares a detection if the buildup in the filter achieves a given level.

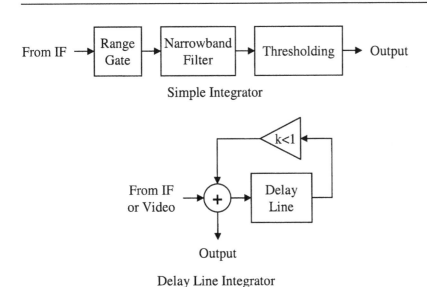

Figure 5.28 Simple integrator and delay line integrator.

The recirculating delay line is an elegant form of integrator that can be used for either pre- or postdetection integration. A rudimentary one is shown in Figure 5.28. The delay of the feedback loop is made equal to the radar's pulse repetition interval. The number of pulses integrated is flexible, making the system potentially useful in phased arrays that can have arbitrary dwell times. The drawback is that k, the loop gain, must be kept slightly below 1 to avoid oscillations, meaning that the integrator falls short of ideal. When used after the envelope detector, the recirculating delay line is suitable for digital processing.

Postdetection integration may also be done by the human eye observing a CRT display. On a relatively noise-free CRT screen, Marcum (1960) found operators able to integrate 30 pulses or less about as well as an optimum post-detection integrator.

5.9.2.2 Doppler processing.

Single-pulse Doppler processing (an approach that is usually practical only on long-range radars) is relatively easy to implement. The radar need only be coherent for the duration of the pulse to be processed. Recall that the length of the pulse is determined by the required Doppler resolution, and the bandwidth of the IF amplifier chain must be sufficient to cover all target range rates.

Pulse burst Doppler processing provides information via a bank of matched narrowband Doppler filters at the output of the final IF amplifier. Each filter matches the radar's pulse burst duration and is tuned to a particular Doppler frequency and thus range rate. The Doppler filters overlap the adjacent filter by about one-half (−3 dB) the filter bandwidth. The Doppler filter bank covers the band of all feasible target Doppler shifts (range rates). Any filter that rings is a target signal that has a particular Doppler shift and thus range rate. If several Doppler filters ring simultaneously, they identify targets at different Doppler shifts (range rates). The decision logic for targets that overlap adjacent filters might be: equal amplitude, assign a Doppler shift (range rate) halfway between the center Doppler (range rate) of the two filters; otherwise, assign Doppler shift (range rate) based on the larger signal amplitude. For a radar dealing with high range rate targets (e.g., aircraft, missiles, satellites, and so forth), the number of Doppler filters can become large.

The use of digital signal processing (with special-purpose programmable microprocessors or microcomputers) in radars has become customary at video frequencies. The ensuing discussion owes much to Stimson, 1998, Section IV, "Pulse Doppler Radar." Because the pulse Doppler waveform (see Figure 5.19) is really a sampling of a very long pulse train, Doppler frequencies appear as pulse-to-pulse modulations. Doppler information can be extracted at video frequencies by mixing down with a reference Doppler signal. A digital pulse Doppler processing flow diagram is shown in Figure 5.29. The quadrature detector (sometimes called a *phase detector* or *synchronous detector*) detects the envelope of the IF frequencies (or a further beaten down version) in two channels. These channels are in phase quadrature (that is, they are 90° apart in phase) and called I and Q (in-phase and quadrature) channels, as given in Equation (5.37). In this way, digital sampling of the signals will never be zero in both channels, the direction phase is advancing (toward the radar or away from it) will always be known, and the angle and magnitude of the resultant are trivial to calculate. The analog-to-digital (A/D) converter samples the analog signals and tags each sample with a binary word telling time and voltage level. By Nyquist's theorem, there must be at least two samples per hertz of the highest frequency present in order not to lose information.

$$I = A \cos\phi \qquad Q = A \sin\phi \qquad\qquad (5.37)$$

where:
I = in-phase phasor of the signal, volts
A = phasor amplitude, volts
ϕ = phasor angle from the in-phase axis, radians or degrees
Q = quadrature phasor of the signal, volts

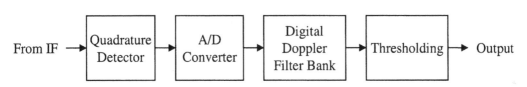

Figure 5.29 Digital Doppler processing flow.

In an analog Doppler processor, the spectral response required of the Doppler filters is known, because the Fourier transform of the transmit pulse is known. To design the digital filter for a pulse Doppler waveform, the Fourier transform of the digitized signal is needed. It is called the *discrete Fourier transform*, and it can be calculated by solving the number of simultaneous equations equal to the number of points in the digitized version of the signal. Efficient ways of doing this calculation for a power of 2 number of pulses (2^{n_p}) are called *fast Fourier transforms (FFTs)* [Sheats, 1977]. However, with the tediousness of the calculation, insights vanish.

A better perspective can be achieved by recalling that all time functions can be reduced to a set of sines and cosines which themselves can be turned into revolving phasors. The phasors have the magnitudes (voltages) and rates of revolution (frequencies) which, when taken together, add up to the time function being represented. The task is to set up in the computer with a number of rotating frames of reference which correspond to the range of Doppler frequencies expected out of the video amplifier divided by the Doppler resolution available. The digitized signal is then processed through these frames of reference. Each passage is the convolution of all the points in the waveform with the particular rotating reference. If the output builds up with the number of samples processed, a Doppler signal is present at that frequency.

The actual calculations that are performed in a digital filter for a single frequency at a single range include two multiplications and an addition in each of the two (*I* and *Q*) channels for every sample, plus four more calculations to RMS the integrated output. For a 16-sample integration, 260 computations are required. If 100 range bins are to be processed and 100 Doppler frequencies resolved in 0.1 sec, a rate of 26 million computations per second is necessary. Considerable parallel processing is often involved. Modern digital processors and microprocessor are up to this task.

Digital processing has many advantages. Once a waveform has been converted into a series of computer words, it can be stored in computer memory, thereby allowing the preservation of amplitude and phase indefinitely. The problems of instability and unreliability in analog circuits are nonexistent with digital processing. Digital processing also has unmatched flexibility. To revise digital processor designs often requires only software modifications.

5.10 EXERCISES

1. A radar transmits an 0.5-μs pulse at a pulse repetition frequency, PRF = 300 Hz. What range resolution and unambiguous range can this radar achieve?

2. A space surveillance radar has the requirement to resolve spacecraft in range if they are 30 m apart and in Doppler if they are moving at range rates that differ by 15 m/sec. Design a compressed pulse at X-band (f_c = 10 GHz) that will meet the requirements.

3. What is the dead zone of the radar whose waveform was designed in the preceding question?

4. A radar approach control radar (maximum range = 40 km, minimum range = 0.5 km) has the requirements to resolve aircraft 30 m apart in range and 15 m/sec separation in range rate.

    A. Why is a compressed pulse unsuitable for the waveform of this radar?

    B. Design a pulse burst waveform for this radar at X-band (f_c = 10 GHz).

5. The waveform of the preceding question is unambiguous in range, but it needs a pulse burst waveform to obtain adequate S/N. Assume aircraft to be handled will have range rates over ±300 m/sec.

    A. Design a waveform that is unambiguous in range rate.

    B. What is the unambiguous range of this waveform?

6. A high-precision C-band (f_c = 3000 MHz) tracker needs to resolve targets 1 km apart in angle, 15 m in range, and 3 m/sec in range rate at ranges of 200 km.

    A. What diameter is the antenna?

    B. What are the characteristics of the waveform?

7. A radar has an FM chirp waveform β = 100 MHz bandwidth over a pulse width τ = 200 μsec.

    A. What is the time-bandwidth product?

    B. What is the anti-jam gain?

    C. What is the range resolution?

    D. What is the Doppler resolution?

8. Show that the S/N after pulse compression processing equals the S/N at the input. Assume FM chirp modulation on the pulse.

5.11 REFERENCES

Barton, D. K., 1979, *Radar System Analysis,* Norwood, MA: Artech House.

Deley, G. W., 1964, *The Representation, Detection Estimation and Design of Radar Signals,* Santa Barbara, CA: Defense Research Corporation.

Deley, G. W., 1970, "Waveform Design," in M. I. Skolnik (Ed.), *Radar Handbook,* New York: McGraw-Hill.

Farnett, E. C., T. B. Howard, and C. H. Stevens, 1970, "Pulse Compression," in M. I. Skolnik (Ed.), *Radar Handbook,* New York: McGraw-Hill.

Goj, W. W., 1993, *Synthetic Aperture Radar And Electronic Warfare,* Norwood, MA: Artech House. One of the best and most concise discussions of the effect of pulse compression and Doppler processing (as one needs for a SAR) on the signal-to-noise ratio.

Jenkins, F. A., and H. E. White, 1976, *Fundamentals of Optics,* New York: McGraw-Hill.

Manasse, R., 1960, *Summary of Maximum Theoretical Accuracy of Radar Measurements,* MI-TRE Tech series Report 2.

Marcum, J. T., 1960, "A Statistical Theory of Target Detection by Pulsed Radar," *IRE Transactions in Information Theory,* Vol. IT-6, Apr., pp. 145–267.

North, D. O., 1943, "An Analysis of Factors Which Determine Signal/Noise Discrimination in Pulsed Carrier Systems," RCA Labs Technical Report PTR6C. (Reprinted in *Proc. IRE,* Vol. 51, July 1963, pp. 1016–1027.)

Rihaczek, A. W., 1997, *Principles of High Resolution Radar,* Palo Alto, CA: Peninsula Press.

Sheats, L., 1977, "Fast Fourier Transform," in E. Brookner (Ed.), *Radar Technology,* Norwood, MA: Artech House.

Skolnik, Merrill I., 2001, *Introduction To Radar Systems,* 3rd ed., New York: McGraw-Hill.

Stimson, George W., 1998, *Introduction To Airborne Radar,* 2nd ed., Raleigh, NC: SciTech Publishing.

Temes, C. L., 1962, "Sidelobe Suppression in a Range-Channel Pulse Compression Radar," *IRE Trans.,* Vol. MIL-6, April, pp. 162–167.

Woodward, P. M, 1955, *Probability and Information Theory with Applications to Radar,* New York: McGraw-Hill.

6

Electronic Countermeasures (ECM)

Because radar finds and measures noncooperative targets, it motivates various countering actions. Motorists might want to deprive the highway patrolman of information about their speed. Certainly, nations want to deny other nations all kinds of information, particularly in times of tension or war. An obvious way of denying radar information is to disrupt the very small amounts of received target signal with spurious energy—either actively, by radiating energy in the radar's frequency band (called *jamming* or *active ECM*), or passively, by dispensing various extraneous objects (called *passive ECM*). Both classes of ECM are aimed at obscuring the target or confusing or misdirecting the radar or its operator. In the United States military, a culture has arisen that is based on the development of ECM and responses to it (called *electronic counter-countermeasures, ECCM*). All types of electronic transmissions, not just radar, are involved. An international professional society, called the Association of Old Crows (AOC), thrives.

Because the fate of nations may depend on the effectiveness of ECM and ECCM, details related to specific hardware are highly classified. Yet, there is a great deal of literature available on ECM and ECCM concepts and techniques unrelated to specific equipment. Van Brunt (1978) discusses about 300 ECM and ECCM concepts. Adamy (2001) provides a "first course" in electronic warfare (EW), while Schleher (1999) provides the "graduate" course. Moreover, there are useful general principles of ECM and ECCM that can be discussed and quantified. Barrage jamming—filling the frequency spectrum in which the radar operates with continuous noise—is particularly amenable to analysis. The jamming waveform can be introduced into the radar antenna through its main-beam or sidelobes, as shown in Figure 6.1. A continuous wave (CW) jammer noise waveform is shown in Figure 6.2.

A measure of effectiveness for a jammer is the signal-to-interference ratio (S/I), which is similar to the signal-to-noise ratio (S/N) discussed in Chapter 1, except that radar receiver thermal noise is combined with jammer noise in the denominator. For the jammer to be effective, its noise must be significantly greater than the receiver thermal noise. Thus, often the measure of jammer effectiveness is the signal-to-jamming ratio (S/J). The ECM designer would strive for S/J < 0 dB, while the ECCM engineer would note that a radar is essentially ineffective when the S/J is less than the detection threshold.

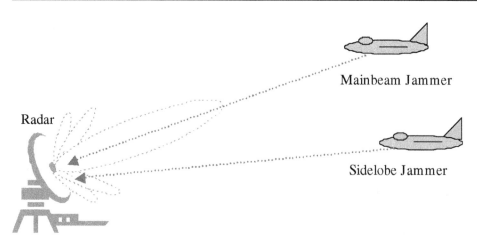

Figure 6.1 An ECM environment.

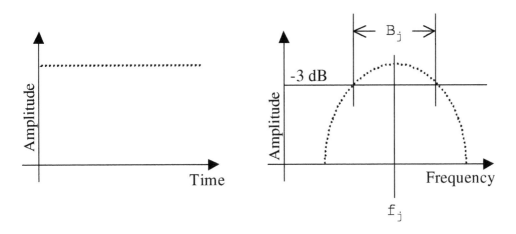

Figure 6.2   Continuous wave jammer noise waveform.

6.1 MAINBEAM JAMMING

A mainbeam jammer is one that transmits jamming noise into the radar's antenna pattern mainbeam. The effect is to obscure target signals within the radar's mainbeam. The jammer noise output fills all range gates and creates a strobe of noise on the radar's PPI scope as the radar beam passes across the jammer's location. Objects along that strobe, which is as wide as the radar's antenna beam, are much harder for the radar to detect. A mainbeam jammer is required to mask only the received target signal in the radar's bandwidth. The ECM community often refers to the received target signal as the "skin return" as it is produced by the reflection off the skin of the target. The jammer might be on a penetrating aircraft or missile, or a more powerful one might be on a standoff aircraft, enabling penetrators to fly toward the radar along a radial from that jammer aircraft.

A noise jammer will have a power-gain product (often called the *effective radiated power*, or *ERP*), as given in Equation (6.1). Sometimes, the ERP is divided by the bandwidth of the jammer noise, giving an effective radiated power spectral density (watt/Hz) or energy. The jammer ERP propagates to the victim radar, resulting in a jammer power density at the radar. If the jammer is on board the radar's intended target, the propagation path is the radar-to-target range, as given in Equation (6.2). The incident jammer power density is collected by the effective area of the radar receive antenna, resulting in the jammer noise at the output of the radar antenna, as given in Equation (6.3). This equation uses antenna gain instead of effective area. Just as with the received target signal, there are losses to the jammer noise power as it propagates to the radar, polarization mismatch with the radar antenna, and in the radar receiver. Incorporating these losses in Equation (6.4) results in the jammer power in the radar receiver (but before detection), as given in Equation (6.4).

$$ERP_j = \frac{P_j G_j}{L_{tj}} \tag{6.1}$$

$$\frac{P_j G_j}{(4\pi) R^2 L_{tj}} \tag{6.2}$$

$$\frac{P_j G_j G \lambda^2}{(4\pi)^2 R^2 L_{tj}} \tag{6.3}$$

$$\frac{P_j G_j G \lambda^2}{(4\pi)^2 R^2 L_j} \tag{6.4}$$

where:
ERP_j = jammer effective radiated power, watts
P_j = jammer peak transmit power, watts
G_j = jammer transmit antenna gain, no units
L_{tj} = jammer transmit loss, no units
R = radar-to-target/jammer range, meters
G = radar receive mainbeam antenna gain, no units
λ = wavelength, meters
L_j = jammer-related losses, no units

The jammer noise power is usually not perfectly matched to the radar waveform (carrier frequency) and receiver bandwidth, as shown in Figure 6.3. The

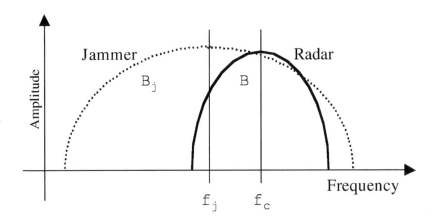

Figure 6.3 Jammer waveform mismatch to the radar waveform and receiver bandwidth.

radar receiver accepts signals only within its bandwidth. Only the portion of the received jammer noise power within the radar receiver bandwidth passes through the radar receiver. The received jammer noise power is reduced by the ratio of the radar receiver bandwidth to the jammer noise bandwidth. The jammer noise power that passes through the radar receiver is given in Equation (6.5). This equation assumes the jammer noise bandwidth is greater than the radar receiver bandwidth, and the center frequency of the jammer noise is approximately equal to the radar carrier frequency.

$$J = \frac{P_j\, G_j\, G \lambda^2}{(4\pi)^2 R^2 L_j}\left(\frac{B}{B_j}\right) \tag{6.5}$$

where:
  J = received jammer noise peak power, watts
  B = radar receiver bandwidth, hertz
  B_j = jammer noise bandwidth, hertz

The frequencies and bandwidths of the jammer and the radar are of vital importance. To extend its bandwidth, thereby forcing the jammer to do the same, the radar may be designed to hop around in frequency, a feature known as *frequency diversity*. The "hopping" may be from pulse burst to pulse burst, pulse to pulse, periodic (fixed or random), or combinations. The radar always knows what frequency it is transmitting. Unless the jammer is combined with a receiver to follow the radar's frequency changes, it is forced to transmit a very wideband noise signal. Frequency diversity gives a measure of "anti-jam gain" or "anti-jam margin." The terms are ambiguous and often include other parameters such as antenna gain, sidelobe levels, power, number of pulses integrated, and so on.

As developed in Chapter 5, a radar using some kind of waveform for pulse compression will have a wide signal bandwidth. It can represent an "anti-jam gain" of the radar over the jammer, because the jammer must cover the wider spectrum of the pulse compression radar signal. For a pulse compression waveform, the radar receiver bandwidth is the modulation bandwidth of the pulse, and this bandwidth is used in Equation (6.5).

Inside the radar receiver there is a complex (magnitude and phase) combination of the received jammer noise and receiver thermal noise power. The result is called the interference signal, as given in Equation (6.6). The signal-to-interference ratio (S/I) is a measure of the strength of the received target signal power relative to the interference signal. The S/I is the ratio of the power from a single received radar pulse to the power of one sample of the interference noise, as given in Equation (6.7). The effect of the integration of multiple pulses on the S/I is given in Equation (6.8). Note: readers can refresh their understanding of the single pulse S/N and associated variables in Chapter 1, and the multiple pulse S/N and associated variables in Chapter 3. The concepts hold true here for the interference signal as well, because it is a noise signal just like the receiver thermal noise.

$$I = N + J$$

$$S/I = \frac{S}{N+J} \tag{6.6}$$

$$S/I = \frac{P G^2 \lambda^2 \sigma}{(4\pi)^3 R^4 L_s \left[(F_n - 1) k T_0 B + \frac{P_j G_j G \lambda^2}{(4\pi)^2 R^2 L_j} \left(\frac{B}{B_j}\right) \right]} \tag{6.7}$$

$$(S/I)_n = \frac{P\,G^2\,\lambda^2\,\sigma\,G_I}{(4\pi)^3 R^4 L_s \left[(F_n - 1)k\,T_0\,B + \dfrac{P_j\,G_j\,G\,\lambda^2}{(4\pi)^2 R^2 L_j}\left(\dfrac{B}{B_j}\right) \right]} \tag{6.8}$$

where:

I = interference noise peak power, watts
N = radar receiver thermal noise peak power, watts
S/I = single pulse signal-to-interference ratio, no units
$(S/I)_n$ = multiple pulse signal-to-interference ratio, no units

Often, the assumption that the jammer noise is much greater than the receiver thermal noise is used. After all, if it were not, the jammer would have little impact on the radar. The equation for the signal-to-jamming ratio (S/J) at the output of the radar receiver from a mainbeam jammer collocated with the target is given in Equation (6.9). The ECM community often uses the jamming-to-signal ratio (J/S) [Equation (6.10)], as it tells them how strong the jamming is relative to the received target signal.

$$(S/J)_n = \frac{P\,G^2\,\lambda^2\,\sigma\,G_I}{(4\pi)^3 R^4 L_s \left(\dfrac{P_j\,G_j\,G\,\lambda^2}{(4\pi)^2 R^2 L_j}\left(\dfrac{B}{B_j}\right) \right)}$$

$$(S/J)_n = \frac{P_j\,G_j\,\sigma\,G_I}{(4\pi)^2 R^2 L_s}\left(\frac{B}{B_j}\right)\frac{L_j}{P_j\,G_j} \tag{6.9}$$

$$(J/S)_n = \frac{P_j\,G_j}{L_j}\left(\frac{B}{B_j}\right)\frac{(4\pi)R^2 L_s}{P\,G\,\sigma\,G_I} \tag{6.10}$$

where:

$(S/J)_n$ = multiple pulse signal-to-jamming ratio, no units
$(J/S)_n$ = multiple pulse jamming-to-signal ratio, no units

The variables P, G, G_I, B, L_s, much of L_j, and B_j are determined by the radar designer. (Even though B_j is a part of the jammer, the radar designer can enforce its value.) The variables P_j, G_j, part of L_j, and σ belong to the opposition (jammer and target). The most sensitive parameter in the equation is range, which strongly favors the jammer's side. The range-squared term tends to dominate the results, even for heroic integration gains and large power apertures. As the student who does the problems at the end of the chapter will find, only a few watts of mainbeam jammer power are sufficient to blind a long-range radar, particularly if the radar cross section of the target is low.

Several tactics for countering noise jamming are available to the radar operator. Clearly, one of the strongest is the integration of multiple pulses. Another is simply to wait until decreasing range to the penetrating target raises the returned signal power. Target detection achieved in the presence of the jammer is called *burnthrough*, and the range at which it occurs is *burnthrough range*. From the point of view of the jammer designer, this would be the *minimum screening range*. Equation (6.11) gives the burnthrough range assuming the jammer noise is much greater than the receiver thermal noise.

$$R_{bt} = \sqrt{\frac{P\,G\sigma\,G_I}{(4\pi)\,SNR_d\,L_s}\left(\frac{B_j}{B}\right)\frac{L_j}{P_j\,G_j}}$$

(6.11)

where:

R_{bt} = burnthrough range, meters

SNR_d = radar detection threshold, no units

If the mainbeam jammer is not collocated with the target, both the range to the jammer and the range to the target enter into the equation. The S/J for a noncollocated mainbeam jammer is given in Equation (6.12). The equation implies that mainbeam jammers standing off (out of harm's way) are practical, which indeed they are.

$$(S/J)_n = \frac{P\,G\sigma\,G_I}{(4\pi)\,R^4\,L_s}\left(\frac{B_j}{B}\right)\frac{R_j^2\,L_j}{P_j\,G_j}$$

(6.12)

where:

R_j = radar-to-jammer range, meters

In addition to the significance of the bandwidth relationships, both main-beam jammer equations emphasize that radar performance against mainbeam jammers is enhanced by high radiated energy and high-gain transmit antennas. For lethal counter-countermeasures, a second radar can locate the jammer by trilateration, and a home-on-jam missile can be launched against it.

6.2 SIDELOBE JAMMING

A successful sidelobe jammer obscures potential target detections with noise to some minimum range. At that range, targets at different angles from the side-lobe jammer can be seen, making the sidelobe jammer a mainbeam jammer until the range is so short that burnthrough occurs. Sidelobe jamming is more dramatic than mainlobe jamming, as it denies angle data to the radar but re-quires much more jammer ERP to overcome the low sidelobe antenna gain. The signal-to-jamming ratio for a sidelobe jammer is given in Equation (6.13). A sidelobe jammer has to overcome not only the jammer-bandwidth-to-radar-bandwidth ratio, but also the ratio of the antenna mainbeam gain to its side-lobe gain. Generally, higher jammer ERP levels are needed to overcome these ratios. Higher jammer ERP levels can be obtained with a high-gain antenna on the jammer that then must be steered at the victim radar.

$$(S/J)_n = \frac{P\, G\, \sigma\, G_I}{(4\pi)\, R^4 L_s} \left(\frac{B_j}{B}\right) \frac{R_j^2\, L_j}{P_j\, G_j\, G_{sl}} \tag{6.13}$$

where:
  G_{sl} = radar receive antenna sidelobe gain in the direction of the jammer, no units

6.2.1 Sidelobe Cancellers

Because the power level in the sidelobes of a well designed antenna is near iso-tropic, the addition of an auxiliary antenna having low gain and wide-angle cov-erage can be used to receive jammer noise independently from the radar antenna. The jammer noise received in the auxiliary antenna is mixed to IF, and its phase is changed by 180°. This phase shifted jammer noise is then used to cancel the noise coming out of the radar receiver. Concepts of coherent side-

lobe cancellers and adaptive ECCM were greatly enriched by the late Paul Howells [Howells, 1976]. Farina (1992) discusses many SLC concepts and their
strengths and weakness. Figure 6.4 is the flowchart of a sidelobe canceller
(SLC) at IF, although Howells and others used mixing rather than summation.
Cancellation might be done by the radar operator, or, more elegantly, by treat-
ing the feedback loop from the canceller output as an error signal to continuously drive the noise to a minimum.

The jammer power arrives at the antenna elements off boresight, so it does
not add up in phase when combined. Power arriving at one element will be out
of phase (sometimes by many wavelengths) with that arriving at another element. In other words, the sidelobes are suppressing the carrier and its modulation. Nevertheless, the sum of the real parts of all these phasors oscillates at the
same rate as the carrier frequency. Thus, coherent cancellation can take place
at RF or IF but will be imperfect. Jammer noise modulation, oscillating much
more slowly than the carrier, will be smeared somewhat as it transits the aperture. The degree of smearing depends on the angle of arrival, the aperture di-

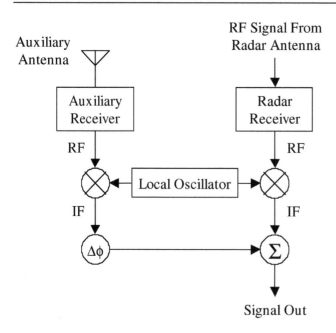

Figure 6.4 Flow diagram of a sidelobe canceller.

mension, and the bandwidth of the radar's IF filters. Even though the envelope of jammer noise has been made deterministic and can be cancelled, the receiver noise in the auxiliary element is independent and random. It adds to that in the radar receiver, doubling the noise power of the system. The increase in thermal noise and the fact that whatever signal is present in the auxiliary receiver is also cancelled serve to provide an upper bound on the effectiveness of a single coherent canceller, as given in Equation (6.14).

$$(S/N)_{pc} = \frac{S - S_{sl}}{2N} = \frac{S}{2N} \qquad (6.14)$$

where:

$(S/N)_{pc}$ = post-cancellation signal-to-noise ratio, no units
S = target signal present in the radar receiver, watts
S_{sl} = target signal present in the auxiliary receiver, watts
N = radar receiver and auxiliary receiver noise (assumed to be the same), watts
S/N = nonjammed signal-to-noise ratio, no units

Because target signal present in the auxiliary receiver is much smaller than the target signal in the radar receiver, it can be ignored. As the number of auxiliary receivers is increased, the noise components continue to add, further reducing the SLC effectiveness. However, multiple auxiliary receivers are needed to cancel multiple jammers. There are other small residuals in the canceling process, such as radar instabilities, filter mismatches, and the SLC response time.

The action of the SLC can also be viewed from the perspective of phased-array antenna design. Implicit in the existence of nulls in antenna sidelobes is the notion of adaptively moving those nulls to the angles of arrival of sidelobe jamming, i.e., "null steering." Recalling the discussion of sidelobes in Chapter 2, the locations of sidelobe minima and maxima depend on the aperture dimensions and the weighting of the signal across the aperture. Changes of phase of particular elements or assigning of phase and amplitude characteristics to auxiliary elements can create nulls in arbitrary locations. Figure 6.5 is a computer simulation of a 21-element array in which a null has been placed in the center of the second sidelobe by the introduction of a phasor of equal amplitude and

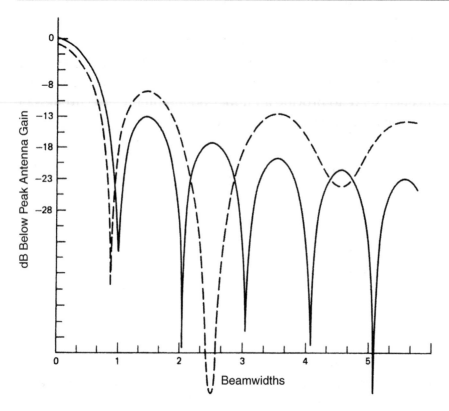

Figure 6.5 Canceling a sidelobe.

180° phase difference. Notice, however, how the overall pattern is modified—mainbeam gain is reduced, and the farther-out sidelobes become higher
(you never get something for nothing). Bear in mind also that these are CW
patterns. A wideband signal will effectively widen the sidelobe nulls slightly,
leaving residues of uncanceled noise.

The cancellation concept can also be applied to the mainlobe. The difference beam in a monopulse tracker (see Chapter 3 for a discussion) does mainbeam cancellation. Pointing a monopulse tracker at a jammer cancels the noise
in the difference beam. Targets slightly separated in angle from the jammer
can be picked up in the split beams of the monopulse system.

Phased arrays, which are designed to respond optimally to a variety of environments, are called *adaptive arrays*. Conceptually, they have great prowess. In

practice, their cost, complexity, intensive processing requirements, and settling time have worked against them. Modern signal processing advances are reducing the processing requirements, although cost and complexity remain.

6.3 LOW PROBABILITY OF INTERCEPT (LPI) RADAR AND RADAR WARNING RECEIVERS (RWRs)

Because radars can be located in angle by their emissions and attacked as a result, there is a continuing need for radars that are not easily detectable. The main idea is to keep the radar off the air as much as possible, emitting only short pulse bursts, each with minimum power, at a different frequency, with a wideband modulation. The radar designer seeks four attributes: narrow beamwidths, ultra-low sidelobes, tailored power output, and a low-power-level noise-like modulated signal. The last of these has become practical only recently because of the computational capacity that digital processing microelectronics have brought us. Wideband spread spectrum signals with pseudorandom noise modulation are a reality. Heretofore, ECM receivers, like the radar warning receiver (RWR) carried in our aircraft, have kept up with the detection and cataloging of radar signals. LPI radars may change that, but not without difficulty. Consider the requirement of an LPI radar: to detect targets at substantial ranges without being detected itself. The RWR has only a one-way propagation path, whereas the radar has two-way (round trips). LPI radars have an R^2 disadvantage to overcome.

From Chapter 1, the single pulse S/N received of the radar is given in Equation (6.15). The single pulse S/N of the RWR (ratio of the received signal power in one radar pulse to one sample of the RWR thermal noise) is given in Equation (6.16). Assuming the S/N for detection (detection threshold) to be equal in both systems, the two equations can be equated, and canceling out like terms yields Equation (6.17). If similar technology is available to both systems (a conservative assumption), then system noise temperatures, noise figures, and losses are the same. With these assumptions, we are left with Equation (6.18).

$$S/N = \frac{P\,G^2\,\lambda^2\,\sigma}{(4\pi)^3 R^4 (F_n - 1)\,k\,T_0\,B\,L_s} \tag{6.15}$$

$$(S/N)_{RWR} = \frac{P\,G\,G_{RWR}\,\lambda^2}{(4\pi)^2 R^2 (F_{RWR}-1)\,k\,T_{RWR}\,B_{RWR}\,L_{RWR}} \qquad (6.16)$$

$$(4\pi)R^2 = \frac{T_{RWR}\,B_{RWR}\,L_{RWR}}{T_0\,B\,L_s}\,\frac{G\,\sigma}{G_{RWR}} \qquad (6.17)$$

$$(4\pi)R^2 = \frac{B_{RWR}}{B}\,\frac{G\,\sigma}{G_{RWR}} \qquad (6.18)$$

where:

$(S/N)_{RWR}$ = RWR single pulse radar signal-to-noise ratio, no units
G_{RWR} = RWR receive antenna gain, no units
F_{RWR} = RWR noise figure, no units
T_{RWR} = RWR system noise temperature, kelvins
B_{RWR} = RWR bandwidth, hertz
L_{RWR} = RWR system losses, no units

To overcome $(4\pi)R^2$ with processing gain and antenna gain against a small RCS target is an imposing task for a radar system. To see how imposing, consider that at a 100-km range, the ratio to overcome is 111 dB; at 200 km, it is 117 dB! For a 1-m^2 target, and a difference in antenna gain of 30 dB, 81 dB of processing gain is required at 100-km range. Postulating a radar with 1 GHz of pseudorandom noise signal bandwidth and 0.01-sec processing (integration) time gives 70 dB of processing gain, leaving the LPI radar 11 dB short.

Various tactical situations might make up for the deficit or worsen it. The RWR's bandwidth will not be so mismatched as to allow the full signal processing gain by the radar; on the other hand, the RWR's bandwidth must be much wider than the radar's if the radar has not yet been detected. Interactions among the RCS, RWR antenna gain, and range are also likely. If the aircraft with the RWR operates within the operational range of the radar, it may be attacked; but if it stands back, it loses some of its range advantage. If the radar is ground based and the RWR is airborne, the RCS will usually be larger than the RWR antenna gain, the RCS of the aircraft being the sum of the antenna RCS and other reflectors on the aircraft. If the radar is airborne and the RWR is ground based, the RWR antenna gain need only be inconspicuous and/or removed to the rear. A large RWR antenna requires multiple simultaneous receive beams or fast-scanning beams, with accompanying algorithms and processing capacity, and a state-of-the-art RWR requires multiple simultaneous processing channels with a variety of filter bandwidths. Whether LPI radars or RWRs, ground-based systems have advantages over airborne systems, because wide bandwidths and fast angle scanning (including the paraphernalia required for them) and large apertures are easier and cheaper on the ground. Big ground-based antennas have a gain advantage of about 20 dB over big airborne antennas, and 30 dB over engagement radar antennas.

The LPI radar designer can employ a narrow fast-scanning beam to force the RWR to see only his sidelobes most of the time. Given a design requirement that the LPI radar be narrowbeam and have low sidelobes, 40 to 50 dB of additional margin favoring the LPI radar may result.

6.4 OTHER JAMMING TECHNIQUES

So far, we have discussed only the classic barrage noise jamming modes. Still to be considered are tactics and techniques that generate false targets or take advantage of specific radar characteristics. In assessing what follows, remember that for every ECM there is an ECCM, and for every ECCM an ECM, ad infinitum. Consequently, there are continuing efforts in peace and war to keep one step ahead of everyone else.

Repeater jammers sample the incident radar signal and return it to the radar. In doing so, they may amplify the signal so that the normal return from the target is masked. They may reradiate it immediately or delay the return. By delaying the return, false target are generated in range. They may store the return and repeat it several times, possibly changing the frequency to simulate different range rates, or alter the delay to simulate changing ranges. ECCMs for repeater jammers include rise-time detection, cross correlations in angle and Doppler, and use of a second radar for trilateration.

Jammers that attack equipment characteristics usually require detailed knowledge of the radar's mechanisms for range, angle, and Doppler tracking. For example, as described in Chapter 3, a radar that uses conical scan for angle tracking is vulnerable to ECM, which frequency sweeps upward or downward across its conical scan frequency or varies, at the conical scan frequency, the amplitudes of target returns. The ECCM for this is to have a variety of conical scan frequencies, scan on receiver only, or switch to monopulse angle tracking. A radar that does range tracking by keeping the centroid of the signal in the center of a tracking gate is vulnerable to a repeater jammer that covers the target's return with a large amplitude signal whose delay is gradually increased, moving the tracking gate away from the true target. This ECM technique is called *range gate pull off (RGPO)* or *range gate stealing (RGS)*. An ECCM for this is tracking algorithms that generate a new track on the large amplitude false target without dropping track on the real target, and dispose of the false target by cross-checking in the Doppler channel. Similarly, a Doppler track can be broken by a repeater jammer that generates a strong signal at the target Doppler and gradually changes its frequency. This ECM technique is called *velocity gate pull off (VGPO)* or *velocity gate stealing (VGS)*. The ECCM is to maintain both Doppler tracks and cross-check in the range channel.

Of course, possible countering actions come to mind for all of the above ECCMs, creating a never-ending cycle. Consequently, a more general approach is advocated by many [Skolnik, 1980, pp. 547–553; Howells, 1976]. Paul Howells, particularly, enunciated a philosophy for coping with all ECM (in principle, even those techniques that had not yet been invented) based on netted high-resolution systems (in angle, range, and Doppler) using adaptive processing.

6.5 PASSIVE ECM

The principal instruments of passive ECM are chaff and decoys. Use of either necessitates a variety of approaches. We include decoys that have active components as a part of passive ECM, an arbitrary choice. The radar characteristics of simple chaff (half-wave dipoles), including peak and average RCS of both a single dipole and a cloud of n dipoles, are treated in Chapter 4. Because dipoles need be only a half-wavelength long and thick enough to maintain their structure, a few pounds can create very large radar cross sections (as in the answer to Exercise 5, Chapter 4: 100,000 L-band dipoles produce an average RCS of 1,400 m^2). Such a large RCS would mask the presence of targets in its vicinity. Chaff is popular because it is light and simple to make, although temperamental to deploy. A cloud may contain several lengths of chaff to be resonant in the frequency bands of several radars. It is used in ground and air operations and as part of the penetration systems of ballistic missiles. In the atmosphere, chaff's lightness causes it to stop quickly and hang in the air, so a Doppler radar can differentiate it from moving targets. In space, however, chaff moves along with the targets it is masking until it encounters the palpable atmosphere.

Chaff can be designed to be retrodirective; that is, to provide gain back toward the radar that is illuminating it. The radar corner reflector discussed in Chapter 4 is retrodirective. Figure 6.6 shows two examples of a retrodirective chaff: eight trihedral corner reflectors nested together and a jack. The RCS of a corner reflector varies with angle but is substantial everywhere. The nested corner reflectors provide 4π steradians angular coverage. The RCS of a trihedral corner reflector was treated in Chapter 4. At X-band ($\lambda = 0.03$ m), a corner reflector with 0.3-m sides gives an RCS of 339 m^2. A simpler chaff-like structure, also providing returns over multiple angles, is the jack. The jack is three-dimensional crossed dipoles. When dispensed in large numbers, retrodirective reflectors can play havoc with a radar, flooding adjoining resolution cells in range, angle, and Doppler sidelobes and saturating the radar processors.

Whereas chaff is used principally for masking targets, decoys are used to simulate them. Decoys can be effective only if they can provide target simulation at weights, complexities, and costs that are much less than for real targets. To decoy a radar requires simulating RCS and range rate, implying a small vehicle with RCS enhancements. In simple decoys, these can take the form of embed-

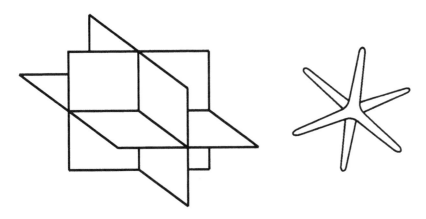

Figure 6.6 Retrodirective reflectors—a 4π corner reflector and a jack.

ded reflectors that have gain in the directions of likely radar viewing. Versions that are more complex could provide electronic enhancements of RCS. Alternatively, decoys might carry the same point-masking repeater jammers as the target. When the target features are changed to make the decoy simulation more effective, the process is called "anti-simulation."

Decoys that accompany aircraft, helicopters, ships, or even ground vehicles on their operational missions are relatively expensive and are operationally complicated. More practically, decoys are carried along, to be dispensed at crucial periods, reducing the time they need to be credible simulators. Such decoys can be free fall, powered, or towed. Decoys that accompany missile warheads may not have operational problems, but the strains on credibility as they sustain exponential decelerations in the atmosphere are severe.

The general ECCM to all passive ECM are the coherent, high-resolution, multifunction radar systems invoked by Paul Howells and others. No ECM can achieve credibility against a radar that has high resolution in all "five dimensions" exploited by radar (three spatial, plus Doppler and time).

6.6 EXERCISES

1. To decide whether receiver system noise needs to be included in the mainbeam jammer equation, calculate the ratio of jammer noise to receiver thermal noise out of the radar receiver. The jammer is at a range R_j = 200 km, power P_j = 10 W, antenna gain G_j = 1 dB, and jammer noise bandwidth B_j = 50 Mhz. The radar has mainbeam antenna gain in G = 40 dB, effective noise temperature T_e = 1500 K, wavelength λ = 0.05 m, and the receiver filter matched to a 1-μsec pulse.

2. The jammer in the preceding exercise is aboard an aircraft flying toward the above radar, which has a peak power P = 1 MW, radar-related losses L_s = 6 dB, and integration gain G_I = 15 dB. What is the S/J if the aircraft has a nose-on RCS σ = 1 m², of 10 m²?

3. Assume that a signal-to-jamming ratio S/J = 6 dB is required for detection. Using the radar and jammer characteristics from the previous two exercises, what is the burnthrough range of the radar for an RCS σ = 1 m², of 10 m²?

4. In a tactical operation, sidelobe jamming is to be used by a jammer aircraft to screen aircraft σ = 1 m² RCS attacking a radar. The attacking aircraft can

launch their missiles at $R = 10$-km range. The jamming aircraft must stand off at range $R_j = 100$ km. The radar to be attacked is known to have an antenna mainbeam gain $G = 30$ dB gain, a peak transmit power $P = 1$ MW, delivering ten 1-μsec pulses as it sweeps across a target, uses coherent integration, and has a matched filter receiver. To be conservative, a 0-dB S/J is required for the jammer to be effective. What effective radiated power (ERP) does the jamming aircraft need if the jammer noise bandwidth $B_j = 20$ Mhz? Assume the radar-related and jammer-related losses are the same, and the radar antenna sidelobe gain is one-half the mainbeam gain.

6.7 REFERENCES

Adamy, David, 2000, *EW 101: A First Course in Electronic Warfare,* Norwood, MA: Artech House. Just what the title says, based on Adamy's long-running column in the AOC *Journal of Electronic Defense* magazine.

Farina, Alfonso, 1992, *Antenna-Based Signal Processing Techniques for Radar Systems,* Norwood, MA: Artech House.

Howells, Paul H., 1976, "Explorations in Fixed and Adaptive Resolution at GE and SURC," *IEEE Trans.,* Vol. AP-24, Sept., pp. 575–584.

Skolnik, Merrill I., 1980, *Introduction To Radar Systems,* 2nd ed., New York: McGraw-Hill. Skolnik discusses ECCM in this edition, not the more recent one.

Schleher, D. C., 1985, *Introduction to Electronic Warfare,* Norwood, MA: Artech House. Clear discussions as well as mathematical analysis of ECM and ECCM.

Schleher, D. C., 1999, *Electronic Warfare in the Information Age,* Norwood, MA: Artech House. Additions and enhancements to Schleher's 1985 book, including numerous MATLAB files.

Van Brunt, L. B., 1978, *Applied ECM,* Vols. I and II, Dunn Luring, VA: EW Engineering, Inc. A myriad of details on ECM and ECCM tactics and techniques.

7

Systems Applications

Radar is pervasive in modern life. Although military applications dominate, at least as far as big complex systems are concerned, nonmilitary applications are legion. Many medium-sized private boats and airplanes are radar equipped, the former for navigation and the latter for obtaining accurate altitude. Commercial airlines and ships are almost all radar equipped. Radars are much used in law enforcement. Traffic officers have them, as do the border patrol and others. The Federal Aviation Administration operates at least 40 large radars and numerous small ones to keep track of the nation's air traffic, with the higher altitudes being under radar control at all times. Many companies and individuals with high-value facilities use radars for security.

Military applications include all of those previously mentioned plus enormously varied additional uses associated with the military missions, such as surveillance, reconnaissance, targeting, and weapons delivery—on land, at sea, and in aerospace.

To recite a litany of these specific mission-oriented radars is probably not as useful as picking out a few illustrative systems to discuss. With the groundwork already laid, we can analyze virtually any radar system. In this chapter, we discuss two classes of systems. First, we will examine systems that are somewhat unconventional when compared to the ones we have been discussing. Over-the-horizon radar, radar altimeters, and ionospheric radars are in this category. Second, we will give attention to three applications that mark the thrust of modern radar technology. These are moving target indication (MTI), pulse-Doppler, and synthetic aperture radars (SARs).

7.1 OVER-THE-HORIZON RADARS

The over-the-horizon (OTH) radars we shall discuss bounce signals off the ionosphere, scatter them from targets at or near the ground, and detect these targets by observing the return path via the ionosphere [Fenster, 1977]. Technically, these are over-the-horizon backscatter (OTH-B) radars. Their operational concept is shown in Figure 7.1.

High-frequency (HF) communications have made use of ionospheric refraction to obtain spectacularly long ranges for most of the twentieth century. Ham radio operators talk to the world with their relatively low-power rigs. Even though the very first radars were in the HF spectrum, this was viewed as a disadvantage. The idea of actually radiating via the ionosphere and detecting the backscatter returning by the same path did not seem credible until coherent processing became practical in the early 1960s. The next two decades saw a slow but steady progress, as digital signal processing has permitted unlimited extension of coherent integration times. OTH-B radars exist, and the United States has deployed systems that will provide warning of an air attack while the attackers are still several hundreds of kilometers away. OTH-B radars are also used for detection of drug smuggling aircraft.

What should interest us, however, is how the principles developed for radars at microwave frequencies apply at frequencies that are two to three or-

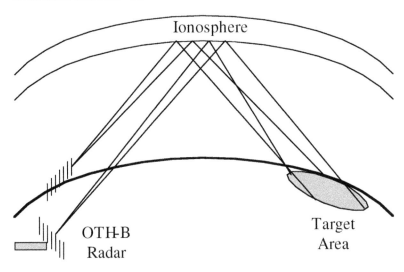

Figure 7.1   Concept of OTH-B radar.

ders of magnitude lower. We shall examine OTH-B antennas, waveforms, and hardware.

7.1.1   Antennas

Because the HF region available for OTH-B radars is about 6 to 30 MHz (the HF region being from 3 to 30 MHz), a 1° beam at the high end of the spectrum would require a 510-m aperture. At 6 MHz, the antenna length would be 2550 m. These antennas are phased arrays. Narrow beams in azimuth are desirable and wide elevation beams are acceptable, which is fortunate for the designer. A 25.5 m antenna height at 30 MHz gives a 20° elevation beam. The resulting antenna looks nothing like a parabolic reflector. In fact, it resembles the antenna farms of major U.S. Navy or Voice of America communications complexes.

OTH-B radar beams are required to propagate out at very low angles to the horizon. This is not a problem at microwave frequencies but is severe at HF where expensive conducting ground planes must be provided in front of the antenna. The electron density of the ionosphere and the available operating frequencies determine at what ranges the radar will operate. Skip distances can vary from several hundred kilometers outward to perhaps 2900 km, bounded by the takeoff angle of the antenna beam, the height of the refracting iono-sphere, and the curvature of the Earth. The radar beam does not radiate into 4π steradians, even conceptually. It is as if the system were operating into a waveguide of infinite, gently curving, parallel planes. Where the one-way signal appears, it may be attenuated by less than $1/R^2$; where it does not exist, of course, attenuation is infinite. Yet, the radar range equation does apply in its classic form to OTH-B radars.

7.1.2 Waveforms

Because the antenna, despite prodigious size, has fairly low gain, and its effi-ciency is not good (its response must cover several octaves of bandwidth), the process of refracting or reflecting from the ionosphere is lossy, and the desired ranges are very long, large amounts of average power are required for success-ful operation. This implies high duty cycle systems and, in fact, encourages 100% duty cycles, with some isolation provided by waveform modulation and the rest by having the receiver at a different location from the transmitter.

To provide adequate surveillance and warning, a good OTH-B system needs good range and Doppler information. Long pulses are no problem, because the minimum range is several hundred kilometers. These factors make a com-pressed waveform of long coherence time but with maximum bandwidth the best choice. Bandwidths are limited to a few percent by physics; the ionosphere reflects differing frequencies to different locations. Although there are coding schemes that can provide equivalent performance, an obvious waveform solu-tion is FM/CW (frequency modulated continuous wave), conceptually identical to the FM chirp waveform analyzed in Chapter 5. An FM/CW waveform is shown in Figure 7.2.

The phenomenology associated with the OTH-B radar guarantees that it looks down on the Earth's clutter. The clutter is a strain on the overall system design, but it becomes a mitigating factor in receiver design (exactly as did mainbeam jamming in Chapter 6). Receiver noise figures and receiver antenna gain become parameters of less concern. The OTH-B radar's ability to reject clutter by range and angle resolution is poor by microwave radar standards. What the OTH-B radar can do is exploit very long coherent processing times to

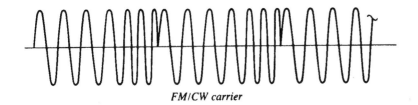

FM/CW carrier

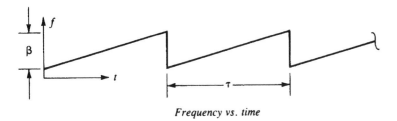

Frequency vs. time

Figure 7.2 FM/CW waveform.

get fine Doppler resolution. The radar system is not the limit here; the coherent processing times of the ionosphere and of the target are. Small wonder the OTH-B designers exploit modern digital signal processing.

Radar cross section is well behaved at HF for targets on the order of a few meters in characteristic dimension. There will not be lobing, scintillation, or other major fluctuations to make the detection statistics difficult. Big transports (Boeing 747s, for example) may be in the resonance region, but their RCS is sufficiently large to make them detectable anyway. However, small private planes or small missiles will be in the Rayleigh region where λ^4 dependency is a quick forecloser. Note, too, that at night, when refraction off the less dense ionosphere necessitates the use of lower frequencies, more targets will sink into the Rayleigh region and become more difficult to detect.

The relatively narrow absolute bandwidths and wide beams make OTH-B inherently poor in anti-jam performance despite extremely narrow band processing. However, given that the jammer is not in physical proximity to the radar, it must find the right spot with respect to skip distance or it will be ineffective. In addition, the OTH-B, by a judicious selection of frequency, may quickly skip elsewhere.

7.1.3   Hardware

Many years of operating world-girdling communications at HF have kept the state of the art moving in antennas; in very high-power, high-duty-cycle, tunable transmitters and in adaptive receivers. This technology is directly applicable to OTH-B. Transmitters of several hundred thousand watts of average power at HF are available. Accompanying ionospheric sounders for keeping track of the ionosphere in real time so that transmitting frequencies can be changed accordingly are also on the market. FM/CW waveform generators and digital signal-processing technology are available from the microwave radar field, as are sophisticated system control and management equipment.

7.2   RADAR ALTIMETERS

Radar altimeters were some of the first commercial applications of radar principles. They transmit energy downward from an aircraft, measure the time until a return is received, convert the time interval to altitude, and display or record it. They differ from big radars in several ways. Because they are relatively short range (≈30 km maximum), their power-aperture product can be small, their receivers noisy, and their signal processing off optimum. Nevertheless, they must have the ability to report altitude to a fine resolution (approximately <30 m) and follow faithfully (in a few milliseconds) rapid changes in altitude—say, the combination of an aircraft in a dive passing over a cliff. Such performance is not too difficult to attain, so the emphasis is on reliability and economy. Most important is minimum encroachment on the carrier vehicle, which means the system must be small, lightweight, and nonintrusive, requiring no changes in vehicle mass properties or aerodynamics.

7.2.1   Antennas

The radar altimeter needs an antenna beam sufficiently wide that altitude readings can be obtained over any specified flight attitude (perhaps a few tens of degrees for transport aircraft to close to 90° for highly maneuverable aircraft) or a beam that is steerable over those angles. The antenna needs to be flush mounted to provide smooth airflow. Horns, crossed dipoles, and slots potted in a dielectric are obvious approaches. Multiple-beam switched arrays are possible and could be made to meet all specifications, but they are generally too complex and expensive to be competitive except in special applications. Although radar altimeter antennas may have gain, they do not have to—the Earth is a big, nearby target.

7.2.2 Radar Cross Section

The target for radar altimeters is the Earth's surface—oceans, mountains, plains, and cities. Most of the time, the radar altimeter beam is near perpendicular to the surface, making the material on clutter coefficients in Chapter 4 essentially inapplicable here. At near-vertical angles of incidence, the clutter reflection coefficient may exceed 10 dBsm over smooth water; may be between 0 and 10 dBsm over smooth, open country; and may be near –10 dBsm over forests or dense undergrowth. Buildings and other man-made structures will provide strong specular returns, and there will be canceling and reinforcing interactions among objects of differing heights, shapes, and electrical characteristics. The resulting reflection coefficient should lie in the region below smooth water and above open country. The area illuminated by a wide beam can be very large—a 20° beam at 10 km illuminates approximately 12.2 km^2. With a reflection coefficient of 1 m^2/m^2, this would be a RCS of 12.2 × 10^6 m^2).

The radar range equation for an altimeter is essentially the same as for clutter, except for RCS. The RCS is not expressible in a single term in the altimeter range equation. It is an integral or summation of all the contributing scatterers. Modifying the clutter radar range equation, Equation (1.14), and Equation (4.28) accordingly, we get Equation (7.1) and Equation (7.2). Combining these two equations results in the altimeter radar range equation, Equation (7.3).

$$S/N = \frac{P G^2 \lambda^2 A_c \Sigma \sigma_i}{(4\pi)^3 R^4 (F_n - 1)\, k\, T_0\, B\, L_s} \tag{7.1}$$

$$A_c = (R\, \theta_{3dB})^2 \tag{7.2}$$

$$S/N = \frac{P G^2 \lambda^2 (R\, \theta_{3dB})^2 \Sigma \sigma_i}{(4\pi)^3 R^4 (F_n - 1) k T_0\, B\, L_s} = \frac{P G^2 \lambda^2 (\theta_{3dB})^2 \Sigma \sigma_i}{(4\pi)^3 R^2 (F_n - 1) k\, T_0\, B\, L_s} \tag{7.3}$$

where:

S/N = single pulse signal-to-noise ratio, no units

$\Sigma\sigma_i$ = sum of the reflection coefficients from all contributing scatters, m^2/m^2

This is a convenient way to view the radar altimeter and can be applicable to narrowbeam, wide-bandwidth systems. However, it does not stand up to detailed design of generalized systems. For example, with a wide beam and a relatively large bandwidth (for fine range resolution), the only RCS of interest is that of first return—which may be quite small. As shown in Figure 7.3, the remaining illumination, spreading out in a set of concentric annular rings from beam center (widths determined by the effective pulse width of the altimeter), is of little interest and does not contribute to system performance.

7.2.3 Waveforms

Not surprisingly, a long-time popular waveform for radar altimeters is FM/CW, as shown in Figure 7.2. The reference ramp can be just the transmitted wave-

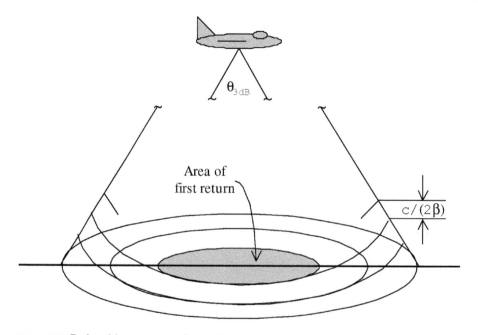

Figure 7.3 Radar altimeter ground return.

form fed into the receiver. FM/CW is simpler and cheaper to implement than pulsed waveforms and places less stringent requirements on amplifiers, switches, power supplies, and cooling.

Several unusual considerations are associated with the design of a suitable FM/CW waveform. One is time sidelobes. The FM modulation has sidelobes in the time domain whose power amplitude is of the $(\sin x/x)^2$ form. This means that, if the detection of the first return is to be made on relatively small RCS (low received signal power), the larger received signal powers (due to larger RCS) that will occur in the succeeding few resolution cells may obscure the first return. Time sidelobes can be suppressed, but at increased complexity. The practical effect, therefore, becomes one of placing a lower bound on the first-return RCS, one that is not related to the system sensitivity.

A second factor is unambiguous range. Ordinarily, the range of maximum altitude would be sufficient to describe this parameter. However, the large RCS of the smooth Earth presents possibilities for ambiguous (second time around) returns. The ambiguous returns may be sufficient to interfere with the first return. Thus, it is prudent to select an unambiguous range greater than the maximum altitude, preferably over the horizon (more on the horizon in Chapter 8).

A third issue is one of resolution and associated accuracy. In the geometry of the system we have described, the ground return becomes an extended target. The measure of range accuracy derived in Chapter 5 tends to locate the "center of mass" of an extended target. This can be a disaster in an altimeter. One solution is to increase the FM bandwidth until the resolution by itself is sufficient to accommodate the accuracy specification.

Lastly, there is an error in range caused if the aircraft has a vertical velocity component. We discussed this in Chapter 5 and found such errors to be substantive in some cases. The design fix is an up–down FM ramp.

7.2.4 Hardware

Currently in operation are excellent FM/CW radar altimeters using less than 10 W of power and having very modest antenna gains and noise figures. At additional expense and complexity, radar altimeters of much improved performance could be designed and produced. There is little general demand for such systems, especially in the modern days of the global positioning system (GPS). Special needs, such as for space vehicles and cruise missiles, can be met on a case-by-case basis.

7.3 IONOSPHERIC RADARS

In the late 1950s and early 1960s, the rapid increases in the power-aperture product of radars created a need to understand the propagating medium better, particularly the ionosphere. As a result, ionospheric radars began to appear. Several located in the arctic regions had the specific objective of understanding the auroral ionosphere, but two located near the equator (one in Peru and the other in Puerto Rico) had more generalized missions of ionospheric physics and radio astronomy. We shall discuss the one in Puerto Rico, known as the Arecibo Ionospheric Observatory [Air Force Office of Scientific Research, 1963].

7.3.1 Mission

The Arecibo radar has the planets of our solar system out to Jupiter within range and can gather information about surface properties (RCS), rotational rates (Doppler), and orbits (range and range rate tracking). In fact, it has made a synthetic aperture radar map of the planet Venus [Rogers and Ingalls, 1970]. However, the radar's primary purpose is to study the Earth's ionosphere. The method is to illuminate the ionosphere (from 50 km to several thousand kilometers altitude) and examine the backscatter. By application of Maxwell's equations to the data in the radar returns, the electron density, the electron temperature, and the strength of the Earth's magnetic field can be inferred.

7.3.2 Parameters

Arecibo's antenna is a 305-m aperture spherical reflector located in a natural bowl. Because it is not a paraboloid, it needs a method of correcting for the spherical aberration. It does this by feeding the antenna with a line feed phased so that the wavelets leaving it add up in the far field of the Arecibo antenna as a plane wave. Reciprocity applies, so we can satisfy ourselves that this is plausible by noting as in Figure 7.4 that, when a plane wave is incident on a spherical cap, all ray tracings cross a line that passes through the sphere's center and is paral-

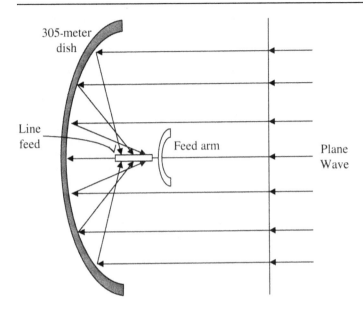

305-meter
dish

Line
feed

Feed arm

Plane
Wave

Figure 7.4 Arecibo spherical antenna.

lel to the plane wave. By placing a line feed in this position, making it an array, and phasing the array to provide the correct phase to each ray, a coherent antenna results. This is how the Arecibo feed works. It has the further attribute of being movable along a feed arm, so the antenna beam can be swung 20° off vertical in any azimuth with little loss in performance. Though not perfect, the antenna is excellent, especially considering the massiveness of its construction. The theoretical diffraction-limited beamwidth ($1.22 \lambda/D$) of the system is 2.8 mrad. The Arecibo antenna achieves 2.9 mrad at zenith.

Because the line feed at Arecibo has high ohmic losses and narrow bandwidth, a new dual reflector feed has been developed at Cornell University with the help of Norwegians [Kildal, et al., 1988.]

Arecibo has a 2.5-MW peak power, 150-kW average power, klystron-driven transmitter. Operating frequency is 430 MHz. Pulse widths of from 2 to 10 μsec, and PRFs from 1 to 1000 Hz, are available. The receiver is a low-noise parametric amplifier. Assuming a 50% efficient antenna, a system noise temperature of 300 K, and system losses of 10 dB, we can quickly calculate (with the tools we learned to use in Chapter 1) that a matched filter receiver for a 10-μsec pulse from the Arecibo radar gives us 10-dB single pulse S/N on a 10-m^2 target at over 19,000 km.

Arecibo, although it has a very special mission, is exemplary of the massive and high-powered radar systems built in the late 1950s and early 1960s. Others of this type include the ionospheric radar in Peru, the extensive ballistic missile early warning system (BMEWS) now being replaced with phased arrays, the frequency diversity air defense radars, and several intelligence-gathering radars.

7.4 ADVANCED TECHNOLOGY RADARS

The availability of coherent circuits, the move to digital electronics, and the rapid evolution of solid-state circuits and devices have greatly expanded the capabilities of radars. A "new generation" of radars has appeared that takes advantage of the new technology. The large family of moving target indicator (MTI), pulse-Doppler, and synthetic aperture radars (which often have esoteric names) belongs to this group. Although they represent extremely complex and intricate applications, their sophistication is achieved by manipulation of the fundamental properties of radars already discussed in the preceding chapters.

7.4.1 Moving Target Indication (MTI)

MTI is really a rudimentary form of pulse-Doppler. Because it has been in operation in various forms since shortly after World War II, it hardly qualifies as "advanced." Yet, MTI radars of the quality now being developed require all that the new technology can provide. Moving target indication does just what its name says: It indicates moving targets while suppressing nonmoving targets.

7.4.1.1 Noncoherent MTI. An early form of MTI was noncoherent. It was called *area MTI* [Barton, 1988, p. 243]. It did not operate pulse to pulse. The returns from one scan were simply subtracted from those of the next. All targets that had moved at least one resolution cell and/or whose amplitudes fluctuated in the time between scans were preserved. All unmoving and/or steady amplitude objects, including fixed clutter, were canceled. The method was practical and still is used on some radars, but it is severely limited in performance.

In this type of noncoherent system, cancellation is done using the envelopes of the returns. Cancellation is never complete because of noise-like factors such as instabilities in the radar, changes in the clutter radar cross section as a function of time, vagaries of the propagating medium, and receiver noise. Whatever was constant about the returns from clutter was canceled, but whatever was noise-like was added on an RMS basis, as given in Equation (7.4). Although this kind of noncoherent MTI presents a relatively clear scope, especially if judicious use of thresholding is employed, digging out a signal that is actually in clutter is problematic. Even assuming two-pulse coherent integration for the target signal, the signal-to-clutter ratio results in Equation (7.5).

$$\sigma_{MTI} = \sqrt{\frac{\varepsilon_1^2 + \varepsilon_2^2 + \ldots + \varepsilon_n^2}{n}} \tag{7.4}$$

$$\frac{S}{C_{MTI}} = \frac{2\sigma}{\sqrt{\dfrac{\varepsilon_1^2 + \varepsilon_2^2 + \ldots + \varepsilon_n^2}{n}}} \tag{7.5}$$

where:

σ_{MTI} = clutter radar cross section residue in a radar cell having clutter but no target, square meters

ε_i = noise-like factors, square meters

n = number of noise-like factors, no units

S/C_{MTI} = signal-to-clutter ratio, no units

σ = target radar cross section, square meters

For fair detection probability, $S/C_{MTI} \approx 20$ is necessary. Clutter radar cross section can be factors of 100 above target radar cross section. To simplify the problem, we can limit it to radar errors. Angle and range errors even on good radars tend to be about one-tenth the range and angle resolution, accounting for a substantial pulse-to-pulse residue from area MTI. Applying these estimates of target radar cross section and range and angle errors, $S/C_{MTI} \approx 0.2$, which is woefully insufficient.

7.4.1.2 Coherent MTI. Coherent MTI takes advantage of the different frequency domain (power spectral density) characteristics of the target signals and clutter signals. Coherent MTI can utilize coherent waveforms or use clutter in the target region as the phase reference to do coherent cancellation in that region [Schleher, 1991]. The different target and clutter power spectral densities (PSD) are shown in Figure 7.5 for a ground-based radar system. If there are no clutter or other targets at different range rates (moving targets) in the target region, there is no output from the MTI. The processing required for this type of MTI, however, is the same as for fully coherent MTI.

Coherent MTI does very well at rejecting clutter of one Doppler shift (or narrow power spectral density). To reject clutter, there are single-delay cancellers, double-delay cancellers, and even triple-delay cancellers. Because the processing is coherent, we can invoke the same approach we used for Doppler processing in the preceding chapter. The frequency response of a two-pulse system (single-delay canceller), as given in Equation (7.6), is shown in Figure 7.6. One null of the frequency domain plot is placed at the Doppler shift of the clutter to be rejected. The other nulls will occur at all Doppler shifts that are positive and negative integer multiples of the PRF.

$$G_p(f) = \left[2\sin\left(\frac{\pi f}{PRF}\right) \right]^2 \qquad (7.6)$$

where:

$G_p(f)$ = signal-delay MTI power frequency response, no units

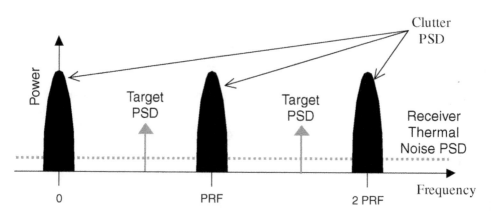

Figure 7.5 Target and clutter power spectral densities.

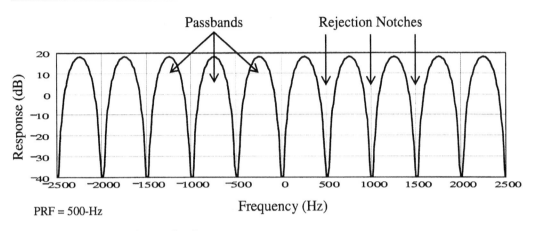

Figure 7.6 Two-pulse MTI canceller frequency response.

The two-pulse canceller does an excellent job of suppressing the clutter at zero Doppler and integer multiples of the PRF. Unfortunately, however, targets near zero range rate and at or near all the "blind speeds" are also suppressed. The blind speed is the target range rate that results in a Doppler shift equal to integer multiples of the PRF, as shown in Equation (7.7).

$$f_d = \frac{2 v_n}{\lambda} = \pm n \, PRF \qquad n = 1, 2, 3 \ldots$$

$$v_n = \pm \frac{n \, \lambda \, PRF}{2}$$

(7.7)

where:

f_d = target Doppler shift, hertz
v_n = nth blind speed, meters/second
λ = wavelength, meters
n = integer number of the PRF, no units
PRF = pulse repetition frequency, hertz

If the depth of the null of the coherent MTI system is considered alone, the clutter rejection of the MTI system is very large. Only radar instabilities contribute to degrading a notch that can be made 80 dB or more deep. The clutter cancellation factor accorded to MTI must take radar, clutter power spectral density, and blind speed factors into consideration. How do we assign the single-delay canceller a figure of merit? If we use the ratio of the targets detected after cancellation to those detected before cancellation, it can be infinite. If we compare the real canceller to an imaginary perfect canceller, it's one-half as good.

Neither of these views is very satisfactory. A better approach (and it is an accepted one) is to ignore the targets and compare the clutter after cancellation to the uncancelled clutter. However, if we assume complicated probability density functions for the clutter, we will be unable to manipulate them well enough to understand the answer. So we use the same approach we took with detection theory in Chapter 3. This is a simplified version of Barton's (1979, pp. 210–217) approach. We can assume that the clutter is noise-like and exponentially distributed above zero Doppler (ignoring for the moment the distribution at negative Doppler shifts), and we can write the clutter voltage as a function of frequency as given in Equation (7.8). We can now express the ratio of the residue of clutter after cancellation to the uncancelled clutter, a cancellation ratio, Equation (7.9), which is a number less than 1. Filling in the functions gives Equation (7.10). Equation (7.10) can be simplified by noting that the approximation given in Equation (7.11) is true for reasonable cases, and making use of our knowledge shown in Equation (7.12) for any meaningful parameters. The resultant clutter cancellation ratio is given in Equation (7.13).

$$C(f) = C_0 \int_0^{PRF/2} \frac{1}{\sigma_c} e^{-f/\sigma_c} df \tag{7.8}$$

$$C_{CR} = \frac{\int G(f) C(f) df}{\int C(f) df} \tag{7.9}$$

$$C_{CR} = \frac{C_0 \int_0^{PRF/2} 2\sin\left(\frac{\pi f}{PRF}\right) \frac{1}{\sigma_c} e^{-f/\sigma_c} df}{C_0 \int_0^{PRF/2} \frac{1}{\sigma_c} e^{-f/\sigma_c} df} \tag{7.10}$$

$$\sin\left(\frac{\pi f}{PRF}\right) \approx \frac{\pi f}{PRF} \tag{7.11}$$

$$e^{-PRF/(2\sigma_c)} \approx 0 \tag{7.12}$$

$$C_{CR} = 2\pi \frac{\sigma_c}{PRF} \tag{7.13}$$

where:

$C(f)$ = clutter voltage, frequency domain
C_0 = clutter amplitude
σ_c = standard deviation of the clutter Doppler
C_{CR} = clutter cancellation ratio
$G(f)$ = MTI voltage frequency response

Equation (7.13) is intuitively appealing, because the ratio of clutter frequency spread to PRF appears directly in the answer. Integrating from 0 to PRF/2 eliminates the half of the function that contains negative range rates. Because it is a mirror image, it does not change the ratio. Neither do any of the other infinite repetitions that occur in the frequency domain. The factor σ_c can be lumped to contain "clutter range rate spread" due to all factors that contribute to less than perfect performance at the output of the canceller. The three most important contributors are radar instabilities, actual range rate spread of the clutter, and the range rate spread introduced by the scanning modulation of the antenna. Radar instabilities can be made very small; range rate spread of clutter tends to be small—less than 1 m/sec for ground clutter and 3 or 4 m/sec for sea clutter. However, the range rate spread from scanning modulation is substantive and places a reliable maximum on the cancellation ratio.

The spread of the range rate components introduced into the system by the scanning antenna is that produced by the radial velocity of the antenna itself. (Note that this modulation factor is present in electronically scanning antennas that use changing phases to steer the beam. There is even spectral content due to a stepping antenna.) As Figure 7.7 shows, the maximum frequency spread is just the difference in range rate of the opposite ends, as given in Equation (7.14). Notice that this is also the frequency the antenna beam scans across a single target, and that single target returns the modulation to the antenna.

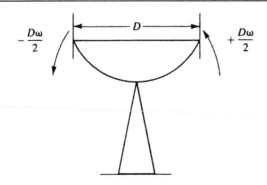

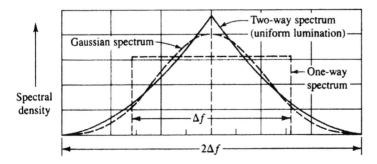

Figure 7.7 Antenna range rate. Copyright © 1976, Artech House, Inc., Norwood, MA, David K. Barton, *Radar Systems Analysis*, p. 215.

$$\Delta f = \frac{D\,\omega}{\lambda} \tag{7.14}$$

where:
  Δf = maximum frequency spread, radians
  D = antenna dimension, meters
  ω = radial velocity of the antenna, radians/second
  λ = wavelength, meters

It can be seen that Δf is uniform across the aperture, that is, it is 0 at the center and changes linearly with distance outward to the antenna edge. The two-way beam voltage is the convolution of the send and receive uniform Δf (or the product of the Fourier transform of that illumination, namely, of form $[\sin x/x]^2$). The two-way beam power is the square of the two-way beam voltage and has extent $2\Delta f$ (see Figure 7.7). Taking the RMS value of the power spectrum of that two-way function gives Equation (7.15). If we assume a normal distribution, we have Equation (7.16). Because *PRF* times T_I gives the number of pulses integrated, plugging σ_c back into the equation for cancellation ratio gives the limit of Equation (7.17).

$$\sigma_c = \frac{f}{\sqrt{10}} = \frac{1}{T_I \sqrt{10}} \tag{7.15}$$

$$\sigma_c = \frac{\Delta f}{3} \tag{7.16}$$

$$C_{CR} = \frac{2\pi}{n_p \sqrt{10}} \tag{7.17}$$

where:
   T_I = dwell time of the antenna on a target, seconds
   n_p = number of pulses integrated, no units

For a 2° beam scanning 360° every 10 sec, and a *PRF* of 1800 Hz, $C_{CR} = -17$ dB, which is far from adequate for severe environments. It is for this reason that double-delay cancellers, blanking circuits, and sidelobe cancellers have been developed and used. The cancellation ratio of a double-delay canceller (two singles in series) may be found by the same method as for a single delay line. The power response of a double-delay line canceller is given in Equation (7.18), and the resultant cancellation ratio is given in Equation (7.19). The power response of single- and double-delay cancellers is shown in Figure 7.8. The power response of an N-delay line canceller is given in Equation (7.20).

$$G_p(f) = \left[2\sin\left(\frac{\pi f}{PRF}\right)\right]^4 \tag{7.18}$$

$$C_{CR} = \sqrt{3}\left(\frac{2\pi\sigma_c}{PRF}\right)^2 \tag{7.19}$$

$$G_p(f) = \left[2\sin\left(\frac{\pi f}{PRF}\right)\right]^{2N} \tag{7.20}$$

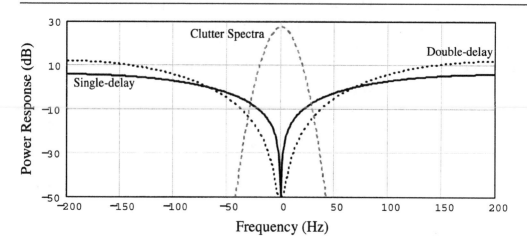

Figure 7.8 Single- and double-delay cancellers.

The coherent MTI feature of deeply suppressing clutter at a particular range rate remains good, but there are too many other limitations. MTI has many blind speeds in the region of range rates of interest, targets at range rates close to the MTI notch are undetectable, and MTI cannot meet the need to suppress clutter at many frequencies rather than one (or, at most, a few). A common solution to the problem of blind speeds is to use staggered PRF MTI processing. With a staggered PRF MTI, an alternating sequence of two or more PRFs is transmitted and processed. The result of using staggered PRFs is the nulls in the MTI response occur at the common frequencies of the PRFs, as shown in Figure 7.9. A staggered PRF MTI allows radar designers to put the first blind speed at the speed associated with the fastest target the radar is designed to detect. The deficiencies of MTI have kept Doppler processing evolving as designers strive to make more complete use of what the physics has to offer. In a sense, advanced MTI detects the clutter so as to suppress it. The new concept is to detect targets, suppressing everything else, but without abandoning MTI cancellers for heavy clutter at fixed range rates.

7.4.2 Pulse-Doppler Radars

The family of pulse-Doppler radars is literally pulse radars that use Doppler information. Doppler processing provides at least three benefits: ability to differ-

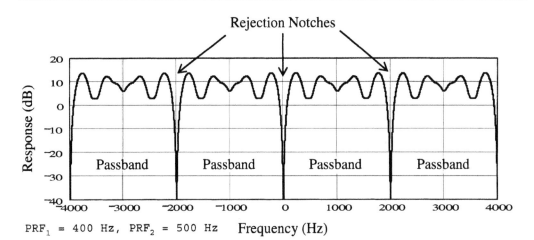

Figure 7.9 Staggered PRF MTI.

entiate between targets on the basis of range rate, improved tracking, and rejection of fixed and low-range rate clutter. From the general waveform analysis provided in the previous chapter, we can quickly develop the principal design features of pulse-Doppler radar waveforms. The pulse width selected will determine the range resolution of the system. The pulse repetition interval will determine the extent of unambiguous range and Doppler, and the dwell time on each target will determine the Doppler resolution. A representative pulse-Doppler waveform was shown in Figure 5.19. The details of the frequency response of the return signals when employed in a ground-based radar are depicted in Figure 7.10. This pattern repeats at each pulse repetition frequency. Immobile or slow-moving targets may be buried in the clutter.

When subclutter visibility is the primary objective (the radar must survey and track relatively slow-moving, close-in targets), the designer will select a waveform with very good range rate resolution. This is to narrow the response of the fixed and low-range rate clutter so that even targets of relatively low range rate will be far out in the clutter sidelobes. The rejection characteristics of the waveform are like $(\sin x/x)^2$, so the first sidelobe is down 13.4 dB and so on.

7.4.2.1 Ambiguity. Most microwave pulse-Doppler radars used for aircraft surveillance over considerable ranges will be ambiguous in either Doppler or

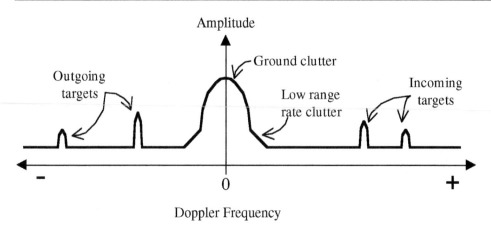

Figure 7.10   Frequency response of the return signals, ground-based radar.

range. The radar designer must decide on whether to have range or Doppler ambiguities. As noted in Chapter 5, both Doppler and range ambiguities can be sorted out by changing the PRF. This sorting algorithm is effective as long as a relatively small number of range or range rate bins is occupied. The radar in the U.S. Air Force's E-3A Airborne Warning and Control System (AWACS) uses waveforms of unambiguous range rate in one of its modes and waveforms of un-ambiguous range in another. Pulse-Doppler radars may also have clutter cancel-lers in their circuitry if the zero-range rate clutter is difficult to accommodate.

7.4.2.2 Airborne pulse-doppler radars. Airborne pulse-Doppler radars are more difficult to design and are more complex than ground-based systems. Consider an aircraft flying over flat terrain of moderate granularity, as shown in Figure 7.11. The region of zero-range rate sidelobe clutter will begin at a spot directly under the aircraft and extend to infinity on each beam of the aircraft. Directly to the front of the aircraft, there will be clutter returns at the speed of the air-craft and, directly to the rear, there will be clutter returns at minus the aircraft speed. In the intervening angles over the compass rose, there will be clutter re-turns ranging from zero to plus and minus the aircraft speed. Clutter range rate with respect to aircraft range rate is given in Equation (7.21). Thus, in range rate space, there is present at all times a band of sidelobe clutter as wide

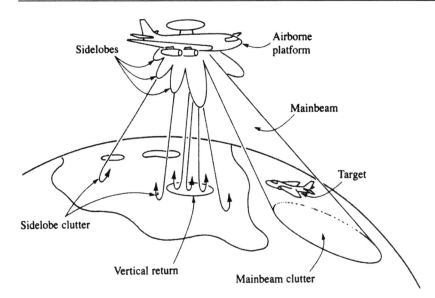

Figure 7.11   Airborne pulse-Doppler radar system.

as twice the aircraft speed. No matter how much or how little of the region the radar mainbeam scans, the sidelobe clutter will be present. The clutter directly beneath the aircraft, because it is specularly reflecting, is particularly troublesome. It may be dealt with by means other than Doppler processing—for example, range blanking or extension of receiver dead time until these returns have passed the aircraft. Both of these techniques will eliminate coverage of close-in targets. If the PRI gives ambiguous ranges, there will be annular rings of range where targets will disappear; if the PRF gives ambiguous range rates, there will be "blind" speeds where targets will be lost.

$$V_{clutter} = V_{ac}\cos\theta \tag{7.21}$$

where:

$V_{clutter}$ = clutter range rate relative to the aircraft, meters/second
$\quad V_{ac}$ = aircraft speed, meters/second
$\quad\ \theta$ = angle from the aircraft velocity vector to the clutter, radians or degrees

The characteristics of Doppler limit the coverage available from an airborne pulse-Doppler radar. The pattern of an airborne pulse-Doppler surveillance radar can be plotted as a function of the range rates and azimuths of possible targets. The wave diagram that results is a comprehensive way of showing surveillance capability in a range rate/angle space that displays the dependence on the range rate of the target and its orientation with respect to the radar-equipped aircraft. An example wave diagram is shown in Figure 7.12. Because it is relatively easy to insert a change in frequency that is a function of the cosine of the change in antenna pointing angle, the airborne pulse-Doppler radar is often made to track mainbeam clutter as having zero range rate—the case in Figure 7.12. Because of the shift of the zero-range rate reference to the mainbeam clutter, the wave diagram appears counterintuitive at first.

First, disregard the target and think of the radar aircraft and its beam. When the beam is pointing straight ahead (0°) and the clutter in the beam is shifted to zero range rate, the sidelobe clutter all around the aircraft is distributed between zero range rate (near the main beam) to minus twice the aircraft speed (directly behind the aircraft). When the beam is pointing at 90° (either side of the aircraft) and the clutter in the main beam is at zero range rate, the sidelobe clutter around the aircraft is distributed in range rate between plus and minus the speed of the aircraft. For the beam pointing to the rear (180°), the zero-degree situation is reversed.

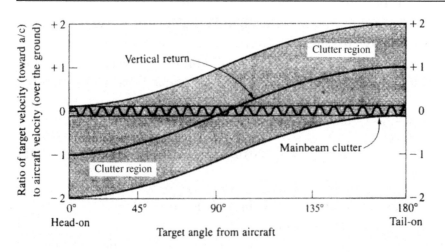

Figure 7.12 Wave map of pulse-Doppler clutter. From *Radar Handbook*, by M. I. Skolnik, © 1970, McGraw-Hill. Used with permission of McGraw-Hill.

Now consider the target. Having mapped the clutter at all radar beam positions, we can be confident that an aircraft in the beam will be out of the clutter only if it is moving at range rates different from the clutter range rates. We know what those range rates are. Thus, to be out of the clutter, a target on the beam of the radar must be moving toward or away from the radar at a speed greater than that of the radar aircraft over the ground. A target in front of the radar will be out of the clutter if it is closing with the radar aircraft at a few meters/second or moving away from it at more than twice its range rate.

Viewed another way, there are two clutter regions in the radar returns; those with zero to positive range rates and those with zero to negative range rates. The altitude (or vertical) return is a zero range rate, and mainbeam return moves with the beam-pointing angle. Figure 7.13 shows this graphically.

Most targets cannot be detected if they are within the width of the Doppler resolution cell for the mainbeam clutter and the altitude return sidelobe clutter. However, targets with range rates of a few meters/second over the Earth may be detectable in that relatively small portion of the sidelobe clutter that is moving at the same range rate as they are. To make such detections requires high-resolution waveforms (in both Doppler and range), high-power transmitters, and very low antenna sidelobes. The antenna of the AWACS airborne pulse-Doppler system has some of the lowest sidelobes ever achieved—including ground-based antennas.

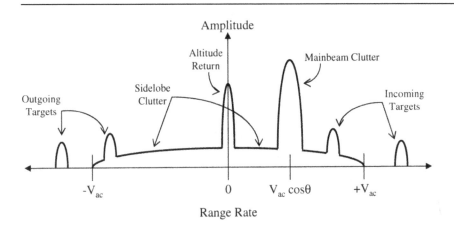

Figure 7.13 Another view of pulse-Doppler clutter, airborne radar.

The airborne pulse-Doppler radar system must be corrected for airplane instabilities, both high frequency (vibration) and low frequency (changes in pitch, roll, and yaw). Conceptually, these corrections are easily made by sensing them with an inertial system and correcting the returns accordingly. Obviously, there are inaccuracies and losses associated with any system of this sort, especially when integration periods can be as high as a few tenths of a second, which is long compared to many aircraft instabilities. Consequently, the accuracy of airborne radars tends to be worse than their ground-based counterparts by factors of ten. Yet their performance is good enough to have revolutionized air battle management in both the U.S. Navy and the U.S. Air Force when the E-2C and the E-3A, both airborne pulse-Doppler systems, became operational.

7.4.3   Synthetic Aperture Radar

The most complex of current radars are the synthetic aperture radars (SARs) [Harger, 1970; Kovaly, 1976; Stimson, 1998]. By measuring received signal strength in a mosaic of range and angle bins, they produce detailed images. They need high resolution in both range and angle. The range resolution of straightforward wideband waveforms is sufficient, but the angle resolution is a difficult matter. The approach to high angle resolution can be viewed in a space-time frame of reference or a frequency frame of reference. In space-time, it can be viewed as using platform motion to provide the physically increased aperture required for high angle resolution. In the frequency frame of reference, the platform motion may be viewed as giving differential Doppler to returns from differing angles; processing these differential Doppler shifts allows high angle resolution.

Most members of the family of synthetic aperture radars are used to make high-resolution ground images. However, because the target can also provide the necessary motion, the SAR concept can be used to image space objects from the ground. This concept is sometimes called "inverse SAR or ISAR." We shall discuss both.

Aside from their obvious complexity and enormous processing requirement (each range and angle cell of a rapidly moving radar beam must be addressed), SARs have several interesting characteristics: the highest resolution is obtained from the smallest physical aperture; the angle resolution obtainable (to first order) is independent of range; if there are any moving targets in the image, they show up in the wrong place; and no matter how low the grazing angle of the radar beam, the image reveals a plan view of the terrain.

7.4.3.1 Imaging ground objects. To discover how SARs work, it is first necessary to show the limitations of the simple unfocused synthetic aperture.

In Chapter 2, we took a series of incremental sources that were all radiating in phase and found that they formed a coherent beam in the far field. We saw that beam formation depended on the relative phases of the incremental sources as they arrived in the far field. We showed later, in discussing phased arrays, that the time when an element radiated was not as important as its phase relationships with the other elements. We can therefore extend this line of thought to state that the far field of an antenna would not be changed if there were only a single incremental element that was placed at each element position in turn and made to radiate with the proper phase—as long as there was an instrument in the far field to store these radiations in their proper phase and add them all up at the end. One small, further step takes us to the situation where the instrument that keeps tabs on the phases and adds them all up properly is at the same location as the incremental radiator, and the path over which this is done is a round trip to a target and back. That is the basis of synthetic aperture. We can give up time to get aperture. (It is not a violation of the first law of thermodynamics, i.e., getting something for nothing, because the time used in synthesizing an aperture is lost to surveillance and tracking that might have been done instead.)

Imagine an aircraft with a fixed beam pointed perpendicular to the line of flight (in other words, "side looking"). Because our system cannot work unless we illuminate a given target each time we move the incremental radiator, our attainable aperture has an upper limit given in the width of the beam of the incremental radiator at some range. A more realistic limit would be the speed of the platform multiplied by the coherent integration time, as given in Equation (7.22). What kind of cross-range resolution can we achieve? We know that the antenna beamwidth is given in Equation (7.23). Thus, the cross-range resolution is given in Equation (7.24).

$$D = V_{ac} T_I \qquad\qquad (7.22)$$

$$\theta_{3dB} \approx \frac{\lambda}{D} \qquad\qquad (7.23)$$

$$\Delta CR = R \frac{\lambda}{D} \qquad\qquad (7.24)$$

where:

D = synthetic aperture length, meters
V_{ac} = aircraft speed, meters/second
T_I = coherent integration time, seconds
θ_{3dB} = antenna beamwidth, radians
λ = wavelength, meters
ΔCR = cross-range resolution, meters
R = radar-to-target range, meters

However, there is a catch: The synthetic aperture length must be short relative to the radar-to-target range. If the synthetic aperture length is an appreciable fraction of the radar-to-target range, the range to a target position will vary slightly over the coherent integration time. Because the wavelength is generally very short, small differences in radar-to-target range can result in considerable differences in the phases of the returns as the synthetic aperture is being formed. The change in phase shift due to range differences is given in Equation (7.25). These phase differences, often called *phase errors*, limit the ability of the returns to sum in the same range bins. The difference in phase also impacts the synthesis of the aperture, just like phase shifts impact an actual array antenna. Using the SAR geometry and phase shift equations, we can derive the maximum effective synthetic aperture length. The synthetic aperture length that results in the maximum antenna gain and minimum beamwidth is given in Equation (7.26). The finest cross-range resolution has been found to be approximately 40% of the maximum effective synthetic aperture length [Equation (7.27)].

$$\Delta\phi = \frac{2\pi\,\Delta R}{\lambda} = \frac{360\,\Delta R}{\lambda} \qquad\qquad (7.25)$$

$$D_{eff} = 1.2\sqrt{\lambda R} \tag{7.26}$$

$$\Delta CR \approx 0.4 D_{eff} = 0.48\sqrt{\lambda R} \tag{7.27}$$

where:
  $\Delta\phi$ = difference in phase, radians or degrees
  ΔR = difference in range (desired range solution), meters
  D_{eff} = maximum effective synthetic aperture length, meters

For a SAR system with a speed of 125 m/sec, integration time of 0.5 sec, wavelength of 0.1 m, and at a range of 20 km, the synthetic aperture is 62.5 m, the maximum effective synthetic aperture is 53.7 m, and the resultant cross-range resolution is 21.5 m, as shown in Equation (7.28). For longer ranges, the cross-range resolution is worse. For flexibility in using a SAR, it would be nice if the cross-range resolution were independent of range.

$$D = V_{ac} T = (125)(0.5) = 62.5 \text{ m}$$

$$D_{eff} = 1.2\sqrt{(0.1)(20 \times 10^3)} = 53.7 \text{ m}$$

$$\Delta CR = 0.4(53.7) = 21.5 \text{ m} \tag{7.28}$$

7.4.3.2 Focused SAR. The limitation on maximum effective synthetic aperture length, and associated cross-range resolution, can be largely removed by "focusing" the synthetic aperture. In principle, to focus a synthetic aperture, all that needs to be done is to apply an appropriated phase correction to the returns received over the duration of the synthetic aperture. Only focused SARs can provide cross resolution that is fine grained enough to allow for recognition of objects having dimensions of approximately 1 m. Returning to our aircraft with its side-looking radar and incremental array, we can treat the problem as an exercise in differential Doppler processing, as shown in Figure 7.14. We postulate a beamwidth for our incremental radiator and assume we can extend the processing time over the whole duration of the beam's transit across the target.

The time it takes the target to transit from the left edge to the right edge of the beam is given in Equation (7.29). The Doppler resolution this system can

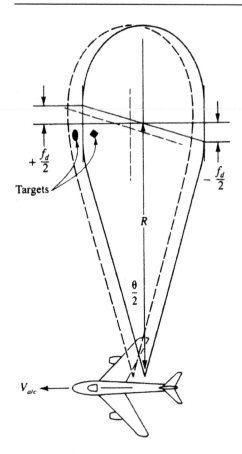

Figure 7.14 Synthetic aperture radar.

achieve is given in Equation (7.30). The maximum Doppler spread available is given in Equation (7.31). This equation uses the small angle approximation for sine, $\sin\theta \approx \theta$ when θ is small and in radians. The number of individual Doppler filters (cross-range resolution cells) obtainable by this system is just the total Doppler spread available divided by the number of hertz in a single Doppler filter (cross-range resolution cell), as given in Equation (7.32). The finest attainable cross-range resolution is the physical extent of the beam divided by the number of Doppler filters, as given in Equation (7.33). This equation shows the cross-range resolution for a focused SAR is independent of range and speed, and that small physical apertures give better resolution.

$$T_t = R\frac{\theta_{3dB}}{V_{ac}} \tag{7.29}$$

$$\Delta f_d = \frac{1}{T_t} \tag{7.30}$$

$$f_{d\ max} = \frac{2V_{ac}\ \theta_{3dB}}{\lambda} \tag{7.31}$$

$$N_f = \frac{f_{d\ max}}{\Delta f_d} = \frac{2R\ \theta_{3dB}^2}{\lambda} \tag{7.32}$$

$$\Delta CR = \frac{R\ \theta_{3dB}}{N_f} = \frac{\lambda}{2\ \theta_{3dB}} \qquad \theta_{3dB} \approx \frac{\lambda}{d}$$

$$\Delta CR \approx \frac{d}{2} \tag{7.33}$$

where:

T_t = time it takes the target to transit from the left edge to the right edge of the beam, seconds

Δf_d = Doppler resolution, hertz

$f_{d\ max}$ = maximum Doppler spread across the spectrum, hertz

N_f = number of Doppler filters, no units

d = physical length of the antenna, meters

These startling outcomes inspired the people at the University of Michigan in the 1950s to initiate their quest for high-resolution airborne SARs.

The reciprocal properties of the SAR aperture are interesting. The resolution obtainable far away in the image is equal to one-half the size of the real aperture on the aircraft, while the lengths of the synthetic apertures created at the aircraft are equal to the beam coverage of the real aperture at any range strip on the image!

7.4.3.3 SAR variations. The SAR is not limited to the creation of strip maps by side-looking perpendicular to the line of movement. It can point at an acute angle (called a *squint angle*) and still create a synthetic aperture. A popular use of the squinted SAR is for *spotlight mapping,* which is valuable for taking navigation fixes and locating targets to attack. In creating a spotlight map, the differential Doppler may not be integrated across the entire real beam. In other words, the SAR may be unfocused or only partially focused, providing what is often called "Doppler beam sharpening." Now, neither a range correction (for spherical aberration) nor a reference Doppler frequency ramp may be required, substantially reducing the processing load.

Doppler integration is limited to the time the returns remain in a range resolution strip. A Doppler filter bank (one filter for each angle resolution cell along a range resolution strip) is sufficient to resolve the differential Doppler. The number of angle cells generated within the real beam is the available Doppler spread divided by the reciprocal of the integration time (Doppler filter bandwidth). As an exercise at the end of the chapter shows, this resolution will depend on the range resolution of the system. It will approximate the unfocused SAR in a nominal case, being much worse than a fully focused SAR.

7.4.3.4 SAR limitations.

Because the SAR uses the Doppler generated by a moving platform surveying a fixed landscape, difficulties arise where there are moving objects on the landscape. In fact, such moving objects, depending on their range rates, may be displaced in angle. (Their ranges not being determined by Doppler, they will remain at the correct ranges.) Because the spectrum of Doppler frequencies available for SAR processing at microwave frequencies and aircraft speeds is not wide, moving ground targets can exceed the ambiguity limits of the waveform and therefore appear at any point on the image. Such wild excursions are not usual, however. A clever interpreter can usually properly replace moving vehicles or trains on their respective roads or railways. However, in a dense target environment where off-road vehicles may be encountered, the problem becomes complicated. Some amelioration is available if the signal is separated, part of it being processed in an MTI channel. In this way, moving targets can be identified and placed. Unfortunately, the accuracy with which they can be placed is not good, because the SAR beamwidth tends to be wide, giving poor instantaneous angle accuracy (range will remain as good as with the SAR processing).

Design of SAR radar waveforms must be accomplished within a rather narrow ambiguity box. To avoid both range and range rate ambiguity, a PRF must be selected to meet the inequality shown in Equation (7.34). The left-hand side of the inequality assumes that the processing interval is between the −3-dB points on the antenna beamwidth. Processing that goes from null-to-null would approximately double the minimum PRF. The right-hand side of the inequality assumes that a clean signal is desired all the way out to maximum range. For a less conservative design, unambiguous range across the width of the strip would be sufficient. If the SAR has a squint angle, the sine of that angle multiplies the left-hand side of the inequality. Notice that long ranges or wide strips and high range rates tend to be incompatible, as do small apertures and long ranges.

$$\frac{2 V_{ac}}{d} \le PRF \le \frac{c}{2R} \qquad (7.34)$$

where:
   V_{ac} = aircraft speed, meters/second
   d = antenna physical length, meters
   c = speed of light, 3×10^8 meters/second
   R = range to the outer edge of the SAR image, meters

For a SAR to work properly, not only are a reference frequency proportional to $2V_{ac}/\lambda \sin\theta$ and a range adjustment proportional to $R(1 - \cos\theta)$ necessary, compensation for the three-dimensional motions of the antenna during flight is also required. This is accomplished by having the equivalent of a three-dimensional inertial measurement unit on the antenna, corrected periodically by navigation sensors (that is, an altimeter and LORAN or the Navstar Global Positioning System).

If the antenna is rigid, corrections for only very limited aircraft motion are practical. If the antenna is movable, either mechanically or electronically, real-time corrections can be made.

All the previously mentioned requirements for data are added to the already prodigious processing associated with measuring received signal strength in every range and angle bin of the radar over several square kilometers of area. Little wonder that, until recently, all SARs had either to record their data, processing it after the flight, or dispatch it directly to the ground by wideband data link. Only at the end of the 1970s had the solid state microelectronic industry advanced far enough to allow image processing and display on board the aircraft in near real time. Space-based SARs are yet to achieve that objective [Elachi, 1982, pp. 57–58].

SAR images can never equal photographs in the visible or infrared for detail (ultimate resolution depends on λ^2), but they are superior in at least two ways: They furnish their own illumination, and they are all-weather operational.

Forming an image based on range and angle bins is the feature of SARs that provides a plan view of the region they are surveying. The derived image emerges as an accumulation of strips at constant range from the platform. Thus, because the grazing angle is not large, there is shadowing but no fore-shortening as there is in a photograph of the same region from the same angle. In fact, the two techniques are reciprocal. In the geometry where a photograph presents a plan view (that is, straight overhead), the SAR would be foreshort-ened to the point of showing only height differences.

7.4.3.5 "Imaging" objects in space.

The first approach to inferring satellite char-acteristics from radar was to take Fourier transforms of the amplitude-time his-tories. This effort was mildly successful but, as better range resolution and digital processing became available, analyzing the phases of the radar returns became popular and successful.

The technique used to obtain angle resolution within a beamwidth in the analysis of satellite signatures has been called by several names: *f analysis, differential Doppler processing,* and *inverse SAR.* Conceptually, it is equivalent to SAR: Use the Doppler spread that occurs as a target changes in angle with respect to the radar line of sight to resolve targets into multiple Doppler bins, with the re-ceived signal strength in each bin belonging at a particular distance from some center of rotation on the object. An assembly of the received signal strengths in their resolved locations plus the range resolution available from wideband waveforms gives a three-dimensional image of the object. This image is seldom pretty, because it arises from coherent microwave illumination rather than dif-fuse light and is at a wavelength five orders of magnitude longer than visible light.

The satellite must meet several conditions to be a good object for signature synthesis.

1. It must have a number of scattering centers; that is, returns must emanate from more than one point on the body. This is an easy criterion to meet; even spheres may have enough irregularities to qualify.

2. The body should be very much larger than a wavelength so as to give detail to the signature.

3. The body must change its attitude with respect to the radar line of sight. The amount of angular rotation required depends on the dimensions of the body.

4. The kinematics of the body must be correctly inferred.

The satellite object identification (SOI) community has done images of many satellites, the national defense applications of which are apparent. One interesting accomplishment has been that, when the National Aeronautics and Space Administration (NASA) was uncertain of the orientation of its Spacelab prior to committing it to reentry, the Massachusetts Institute of Technology's Lincoln Laboratory SOI scientists were able to determine the attitude from their radar data.

7.4.4   Laser Radars

Laser is an acronym for "light amplification by stimulated emission of radiation." The first laser was built in 1960 using a ruby rod [*Encyclopedia Britannica,* 1984]. Since then, however, many materials have been caused to lase. As has been done with microwave devices for many years, the laser generates and radiates coherent electromagnetic energy. The principle of stimulated emission was recognized by Einstein and is implicit in the photoelectric equation. When electrons that have absorbed enough energy to jump to higher orbits subsequently return to lower orbits, they emit photons of a particular frequency, depending on the energy difference between the orbits. Photons introduced into a medium (for instance, a tube containing a gas whose electrons have been excited) will stimulate a shower of photons as the electrons return to lower orbits. If a fully reflecting mirror is placed at one end of the tube at an integer number of wavelengths from a partially reflecting mirror at the other end of the tube, the system will radiate an intense coherent beam.

When a laser is augmented with a receiver and a clock, it becomes a laser radar (ladar), or light detection and ranging (lidar). Lidars are extensively used for making precise measurements of distance. For example, the distance from the Earth to the Moon has been measured to an accuracy of 1 ft by bouncing a laser signal off a reflector placed on the Moon by an astronaut. Other lidar applications include aircraft altimeters, measurers of atmospheric densities and currents, and vehicle speed measurement by law enforcement. The lidar can render high-resolution, three-dimensional images of objects using short pulses and the antenna's real beam. A challenging potential application is for strategic defense—a space-based lidar to provide long-range imaging of attacking missiles in midcourse.

The lidar has many advantages over conventional microwave radar, as well as some shortcomings. Because wavelengths at optical frequencies are on the order of 100,000 times shorter than microwaves, all of the lidar features that depend on wavelength are greatly enhanced particularly aperture gain, measurement resolutions, waveform bandwidth, and target Doppler. As you can calculate using knowledge gained from this book, a 44-dB gain aperture, which required a 15-m diameter antenna at L-band (λ = 0.6 m), is only 0.024 mm across at the frequency of visible light (λ = 0.5 µm). Whereas bandwidths of 500 MHz are hard to come by at microwave frequencies, in the visible range, even a 1% bandwidth gives 600 GHz. Because Doppler frequencies generated by even slowly moving targets are large, lidars do not need complex pulse-Doppler waveforms to do Doppler processing. Single short pulses are usually more than adequate.

Disadvantages are not trivial. Lidars have limited ability to penetrate rain, smoke, haze, and dust, and they are severely attenuated in clouds and fog. Their extremely narrow beams make them poor in surveillance applications (a 0.1-m aperture lidar at a wavelength of 10 µm has a half-power beamwidth of 5.1×10^{-3} degrees). In fact, even when fine angle and range information are provided by a radar, acquisition and tracking of targets, particularly multiple targets, are challenging.

7.4.4.1 Lidar range equation. Because light is electromagnetic radiation, the radar range equation also applies to lidars. Many of the terms are identical, although translations may be required, because lidars are in the domain of physicists specializing in optics, whereas radars are in the province of electrical engineers. Lasers are not spoken of in terms of power-aperture product or effective radiated power (power times antenna gain). Instead, the term *brightness* is used. Brightness is defined [Weiner, 1984] in Equation (7.35). The power intensity at any range can be calculated using Equation (7.36). These expressions are practical, because the narrow laser beam may not subtend the entire target.

$$\frac{P}{\theta^2} \quad \text{or} \quad \frac{PA_e}{\lambda^2 (4\pi) R^2} \tag{7.35}$$

$$\frac{P}{\theta^2 (4\pi) R^2} \quad \text{or} \quad \frac{PA_e}{\lambda^2 (4\pi) R^2} \tag{7.36}$$

where:

P = power, watts
θ^2 = effective aperture beam area, square radians
A_e = effective aperture area, square meters
λ = wavelength, meters
R = lidar-to-target range, meters

The receive antenna of a lidar usually will not be shared with the laser transmitter. A receive antenna has much less stringent tolerances, needing only to focus the incoming radiation in a spot the size of the photosensitive detectors—a square millimeter or so.

7.4.4.2 Lidar S/N. The two terms where the lidar range equation deviates substantially from radar is in detection and radar cross section. At lidar frequencies, electromagnetic energy is more easily treated as quanta (photons) than as waves. In enunciating the quantum theory, Max Planck noted that the energy in a photon was equal to Planck's constant times the frequency of the light. A lidar receiver such as a photomultiplier tube (PMT) simply counts photons during the lidar pulse width (as a matched filter responds to the incoming radar signal). The average number of signal photons arriving at the output of an optical receiver is given in Equation (7.37).

$$\bar{n}_s = \frac{\eta P_r}{h f} \tag{7.37}$$

where:

$\bar{n}_s$ = average number of signal photons arriving at the output of an optical receiver, no units

η = quantum efficiency, watt-seconds or joules

h = Planck's constant, 6.6×10^{-34} joule-seconds

f = frequency of the laser light, hertz

Finding the noise term is more complicated than it is in radar. Thermal noise is swamped by photon noise beginning at frequencies around 10,000 GHz (λ = 30 μm). How much photon noise is present? Just as with radar, the total noise at the output of an optical receiver is increased by sources outside the receiver, including the stars and planets, other stray light, and front-end losses. Nevertheless, the principal contributors are in the receiver, and we will use them to develop an expression for signal-to-noise to compare with radar [Dishington et al., 1963].

Unlike a radar receiver, the noise generated in an optical receiver is increased by the presence of a signal. Because photons arrive randomly, they obey Poisson statistics (a variation of the binomial distribution in which the chances of success in any given trial are very low). The power signal-to-noise ratio for the optical receiver is given in Equation (7.38). Substituting in the received power, and assuming that signal power will be much larger than noise power in cases of interest, gives Equation (7.39). Making the radar and lidar pulse durations equal (and thereby fixing the bandwidth), the expression on the right can be substituted in the radar range equation, as given in Equation (7.40).

$$S/N = \left(\frac{\bar{n}\,\eta}{\sqrt{\bar{n}_s\,\eta}} \right)^2 \tag{7.38}$$

$$S/N = \left(\frac{\dfrac{\eta\,P_r}{h f}}{\sqrt{\dfrac{\eta\,P_r}{h f}}} \right)^2 = \frac{\eta P_r}{h f} \tag{7.39}$$

$$S/N = \frac{P_r}{k\,T_s} \tag{7.40}$$

where:

S/N = signal-to-noise ratio, no units

$\bar{n}$ = variance of fluctuations in the Poisson distribution, no units

k = Boltzmann's constant, 1.38×10^{-23} joule/kelvins

T_s = receiver system noise temperature, kelvins

For a comparison of radar versus lidar, take the quotient (noise term) and insert some values. In the case of a microwave radar receiver at 300 K and a blue-green laser ($\lambda = 0.5$ μm), the radar would have approximately 96 times less noise, as shown in Equation (7.41).

$$k\,T_s = (1.38 \times 10^{-23})(300) = 4.14 \times 10^{-21} \text{ J}$$

$$h\,f = (6.6 \times 10^{-34})\left(\frac{3 \times 10^8}{0.5 \times 10^{-6}}\right) = 396 \times 10^{-21} \text{ J}$$

$$\tag{7.41}$$

7.4.4.3 Lidar RCS. The differences between lidar and radar in radar cross-section calculations arise from the wavelength dependence of RCS. The microscopic granularity of the reflecting surfaces is a powerful factor. In addition, objects in the resonant (or Mie) region and the Rayleigh region are molecular in size.

Because the real beam of a lidar is so narrow, real beam images with good resolution are feasible. However, because the size of surfaces that give specular reflections is so small, large amounts of high cross-section returns (called *speckle*) occur, often suffusing the image. Special limiting circuitry is used to suppress speckle.

The general equation for lidar RCS is the same as that used in radar [Equation (7.42)]. However, in optics, a term for the reflectivity of the target is inserted [Jelalian, 1977]. Adding in the expression for gain gives Equation (7.43).

$$\sigma = G\,A \tag{7.42}$$

$$\sigma = \rho \frac{4\pi \, A^2}{\lambda^2} \qquad\qquad (7.43)$$

where:

σ = radar cross section, square meters
G = gain back in the direction of the lidar, no units
A = effective intercept area of the target, square meters
ρ = reflectivity of the target to the lidar wavelength, no units

For an isotropic scattering target, such as a perfectly conducting sphere, the radar engineer will set the gain of the target to unity; the optical physicist will substitute the gain of a diffuse optical (Lambertian) scatterer, which is 4. Thus, the RCS at laser frequencies of a diffuse target is given in Equation (7.44).

$$\sigma = 4\rho \, A \qquad\qquad (7.44)$$

The equations for corner reflectors at optical frequencies, given a reflectivity of 1, are the same as for radar corner reflectors, but they are physically small for very high RCS (see exercise at the end of this chapter).

When a lidar's beam is so narrow that it does not illuminate the entire target, the RCS is for an "extended target," which is easily calculated, because the intercept area is the solid angle of the transmit beam multiplied by the square of the range to the target as given in Equation (7.45). This equation assumes far-field illumination and a circular beam. Thus, the RCS for a partially illuminated target is given in Equation (7.46). When this equation is put into the radar range equation, the power arriving back at the receiver becomes only range-squared dependent, as with the radar altimeter.

$$A = \frac{\pi}{4}(R\,\theta)^2 \qquad\qquad (7.45)$$

$$\sigma = \pi\rho(R\,\theta)^2 \qquad\qquad (7.46)$$

7.4.5 Bistatic Radar Systems

A bistatic radar system is one where the transmitter and receiver are separated by a significant distance (Blake, 1986; Schleher, 1991; Skolnik, 1980, pp. 553–560). The separate transmitter, receiver, and target form a triangle, as shown in Figure 7.15. Bistatic radar systems offer economic and operational advantages over monostatic radar systems; several receivers may share a single transmitter, and in being passive, the receivers can be difficult to locate. Traditional bistatic radar work has focused on using cooperative radar transmissions. Just like monostatic radar systems, bistatic radar systems can detect and measure targets in the range, Doppler, and angle domains. Depending on the signal amplitude and modulation used by the transmission source, one domain may be better (in terms of longer detection range, finer measurement resolution, and/or acceptable measurement ambiguities) than the others. A few experimental bistatic radar systems are currently demonstrating detection ranges of a few hundred kilometers (km), which is plenty adequate for most air surveillance applications. Developing accurate target position tracks at comparable ranges from these detections is often the challenge.

7.4.5.1 Bistatic range equation.   We can develop the bistatic radar range equation much the same way we developed the monostatic radar range equation in Chapter 1. The transmitter directs the radar waveform at the target. The transmitted waveform propagates to the target and results in a power density at the target. The target intercepts this power density and reflects a portion of it to the receiver. The reflected power propagates to the receiver and results in a power density at the receiver antenna. The receive antenna effective area intercepts the received power density. The received target signal power is passed

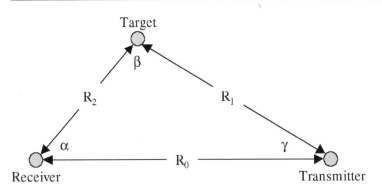

Figure 7.15   Bistatic radar triangle.

along to the receiver. In the receiver, the received target signal power is processed. We can define the ratio of the processed target signal to the receiver thermal noise power, as given in Equation (7.47).

$$(S/N)_p = \frac{P\,G_t\,G_r\,\lambda^2\,\sigma_b\,G_p}{(4\pi)^3\,R_1^2\,R_2^2\,(F_n - 1)\,k\,T_0\,B\,L_s} \qquad (7.47)$$

where:

$(S/N)_p$ = processed signal-to-noise ratio, no units
P = peak transmit power, watts
G_t = transmit antenna gain in the direction of the target, no units
G_r = receive antenna gain in the direction of the target, no units
σ_b = target bistatic radar cross section, square meters
G_p = signal processing gain, no units
R_1 = transmitter-to-target range, meters
R_2 = receiver-to-target range, meters
F_n = receiver noise figure, no units
k = Boltzmann's constant, 1.38×10^{-23} watt-seconds/kelvin
T_0 = receiver standard temperature (usually room temperature, 290 K)
B = receiver filter bandwidth, hertz
L_s = bistatic radar system losses, no units

Bistatic radar systems do not have a detection range *per se*; rather, they have a bistatic detection range product. The bistatic radar range equation can be solved for bistatic detection range product as given in Equation (7.48). Bistatic radar detection coverage is a function of not only the bistatic detection range product but also the receiver-transmitter separation. Bistatic radar detection coverage is often described as *ovals of Cassini*. Figure 7.16 shows example bistatic radar detection coverage for different transmitter-receiver separations.

$$R_1 R_2 = \sqrt{\frac{P\,G_t\,G_r\,\lambda^2\,\sigma_b\,G_p}{(4\pi)^3\,SNR_d\,(F_n - 1)\,k\,T_0\,B\,L_s}} \qquad (7.48)$$

where:

$R_1 R_2$ = bistatic detection range product, square meters
SNR_d = signal-to-noise ratio required for detection (detection threshold), no units

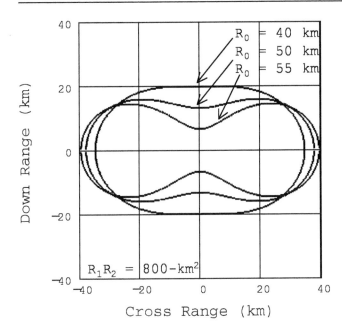

Figure 7.16 Bistatic radar detection coverage—ovals of Cassini.

7.4.5.2 Bistatic RCS.

The bistatic RCS is a measure of the electromagnetic energy intercepted from the transmitter and reradiated to the receiver at the same wavelength by any object. Using the concepts defined in Chapter 4, bistatic RCS is forward scatter. There are many rules of thumb concerning bistatic RCS, many of them trying to relate the better understood monostatic RCS to bistatic RCS. When the bistatic angle is small, the bistatic RCS is approximately equal to the monostatic RCS. Siegel, et al. (1968) report that, for simple, smooth shapes, the bistatic RCS is equal to the monostatic RCS at one-half the bistatic angle. Thus, if the bistatic angle is 50°, the bistatic RCS is equal to the monostatic RCS at 25°. When the bistatic angle is greater than 90°, a commonly held belief is the bistatic RCS becomes significantly larger than the monostatic RCS. For bistatic angles greater than 90°, the target produces forward scatter. While this belief has been shown to be true for smooth, simple shapes, it does not necessary translate to complex targets.

RCS predication codes that address bistatic RCS can be used to determine which rule of thumb is applicable, if any. RCS is a classic example of conservation of energy; we cannot get something for nothing. While we may get a bistatic RCS that is greater than a monostatic RCS for some bistatic angles, there are also some bistatic angles that result in the opposite.

7.4.5.3 Passive coherent location (PCL) bistatic radar systems. Traditional bistatic radar work has focused on using cooperative radar transmissions. Recently, the emphasis has shifted to commercial transmissions, e.g., analog frequency modulated (FM) radio, analog television (TV), cell phones, digital TV, and so on. Extensive digital signal processing is used to extract target echoes produced by reflections from commercial coherent transmissions. Passive coherent location (PCL) is the name for such bistatic radar systems. Just like monostatic radar systems, PCL systems can detect and measure targets in the range, Doppler, and angle domains. Depending on the signal amplitude and modulation used by the transmission source, one domain may be better (in terms of longer detection range, finer measurement resolution, and/or acceptable measurement ambiguities) than the others. The vast majority of PCL systems make extensive use of Doppler processing. A few experimental PCL systems are currently demonstrating detection ranges of a few hundred kilometers, which is plenty adequate for most air surveillance applications. Developing accurate target position tracks at comparable ranges from these detections is often the challenge.

An example of a PCL bistatic radar system is the Manastash Ridge Radar (MRR), developed and used by University of Washington Radar Remote Sensing Laboratory for meteor detection [University of Washington website]. The MRR uses the United States FM radio signal as the transmitter. The MRR receiver system provides a range-Doppler map of the received signal. The range-Doppler map is capable of processing targets up to 350 km. The MRR performs signal processing (combination of coherent Doppler processing and noncoherent averaging) to achieve a maximum Doppler resolution of 1.5 m/sec. The signal processing gain is due to coherent Doppler processing and thus is equal to the time-bandwidth product. The MRR performs Doppler processing of 0.25 sec of data with a signal bandwidth of 20 kHz, which results in a coherent processing gain of approximately 37 dB. The MRR then noncoherently averages 10 sec worth of these quarter-second spectra, or 40 spectra. The resultant noncoherent processing gain is the square root of 40, or approximately 8 dB. The net processing gain for the MRR is 45 dB.

Another example of a PCL bistatic radar system is one built by Paul Howland [Howland, 1997] using the European analog TV signal as the transmitter. This PCL receiver system provides measurements of the Doppler shift and direction of arrival (DOA) of reflections from the TV video carrier. Signals are processed using the fast Fourier transform (FFT) to provide Doppler and DOA measurements. Each FFT requires one to two seconds worth of data. A constant false alarm rate (CFAR) detection scheme is used to identify target reflections and reject noise and unwanted video carrier harmonics. The signal processing gain is accounted for by the narrow receiver noise bandwidth (Doppler processing using an FFT). The resultant range rate measurement resolution is 0.375 m/sec. This PCL bistatic radar detected airliner targets at ranges of several hundreds of kilometers.

7.5 EXERCISES

1. A. An over-the-horizon backscatter radar has its receiver facility located 40 km from the transmitter facility and at right angles to the transmit antenna's boresight. The transmit and receive beams scan 30° simultaneously. Both are unweighted apertures of 20 dB gain in azimuth and 6 dB in elevation at 30 MHz. At that frequency, what is the approximate isolation afforded by the geographic separation of the two facilities? [Hint: Consider a flat Earth. Assume the sidelobe antenna gain is 1/100 (−20 dB) of the main-beam antenna gain.]

B. The system employs an unweighted FM/CW waveform using an FM ramp of 10 µsec duration with a bandwidth of β = 1 MHz. The minimum range is 740 km. What additional isolation is afforded by this waveform? (Hint: Assume waveform time sidelobes are $[\sin(x/x)]^2$.)

C. If the effective radiated power of the OTH-B system were 80 dBW, what signal level direct from the transmitter would enter the receiver front end?

D. Would appear in the first range bin (740 km) at the output of the FM/CW processor?

E. A 10-m^2 target at 740 km range would enter the receiver at -133 dBW, assuming $1/R^4$ attenuation, and would appear at the processor output at about 10 dB below the transmitter power. What are your comments about the need for additional suppression of the direct transmitter signal?

2. A. Design a pulse-Doppler waveform for a transmitted frequency $f_c = 3000$ MHz that is unambiguous in range rate for aircraft flying at 900 m/sec or less (positive range rates only), can resolve aircraft separated by 30 m in range and/or 15 m/sec in range rate.

   B. What is the range ambiguity?

   C. Will the waveform from part (A) support the required measurement performance at UHF ($\lambda = 0.6$ m)?

3. For pulse-Doppler radars on fast-moving aircraft, the clutter problems are complex. Qualitatively evaluate clutter from a hovering helicopter with such a radar.

4. A focused SAR radar, traveling at velocity V_{ac} with plenty of sensitivity, has a λ/d radian beamwidth.

   A. Calculate the Doppler spread across the beam, $f_{d\ max}$, at range R and at range $2R$.

   B. What is the time it takes for the beam to transit a target location, T_p, at those two ranges?

   C. How many cross-range resolution cells (Doppler filters) at R and at $2R$?

   D. What is the cross-range resolution, ΔCR, at R? At $2R$?

5. What generalities can be drawn from the answers to Exercise 4?

6. A focused SAR radar maps a 16 km wide by 16 km long swath in $T_t = 2$ min. With a pulse bandwidth $\beta = 100$ MHz and a real beamwidth of $\theta_{3dB} = 2$ degrees at $f_c = 3000$ MHz, what would be the number of resolution cells processed per second?

7. What is the angular resolution of an L-band ($\lambda = 0.3$ m) "spotlight mapper" squinting at 45°, having a real beam of 2 radians, and with a range resolution of 15 m? (Hint: The time the returns spend in one resolution cell determines the integration time.) Compare the answer to a simple unfocused SAR with the same integration time.

8. A stabilized satellite object is tracked by an SOI radar at UHF ($\lambda = 0.6$ m) for 10 sec. During that time, two RCS phase centers in adjacent Doppler bins are resolved. Amplitude-time histories show that the object rotates in the plane of the radar line of sight every 200 sec. Make a preliminary estimate of the width of the body.

9. What is the projected diameter of an optical corner reflector (cat's eye) with an RCS of 1,000,000 m^2? (Assume that the reflectivity $\rho = 1$, and use a wavelength of 0.5 µm.)

10. A bistatic radar uses an FM radio for a transmitter; peak power $P = 50$ kW, transmit antenna gain $G_t = 0$ dB, and frequency $f_c = 100$ MHz. The receiver has the following characteristics: receive antenna gain $G_r = 10$ dB, signal processing gain $G_p = 15$, noise figure $F_n = 6$ dB, receiver temperature $T_0 = 290$ K, bandwidth $B = 200$ kHz, and losses $L_s = 7$ dB. If the detection threshold $SNR_d = 13$ dB, what is the bistatic range product for a target bistatic radar cross section $\sigma_b = 5$ m^2?

7.6   REFERENCES

Air Force Office of Scientific Research, 1963, *Arecibo Ionospheric Observatory*, Arecibo, Puerto Rico.

Barton, D. K., 1979, *Radar Systems Analysis*, Norwood, MA: Artech House.

Barton, D. K., 1988, *Modern Radar System Analysis*, Norwood, MA: Artech House.

Blake, Lamont V., 1986, *Radar Range-Performance Analysis*, Norwood, MA: Artech House.

Dishington, R., W. Hook, and R. Brooks, 1963, "The Performance of RF and Laser Radar Systems," Report 9200.2-137, Los Angeles: TRW Space Technology Laboratories.

Encyclopedia Britannica, 15th ed., 1984, "Lasers and Masers," Vol. 10, pp. 686–689, Chicago: Benton.

Elachi, C., 1982, "Radar Images of the Earth from Space," *Scientific American*, Vol. 247, no. 6, December 1982, pp. 54–61.

Fenster, W., 1977, "The Application, Design, and Performance of Over-the-Horizon Radar in the HF Band," *Proc. International Conference, Radar-77*, pp. 36–40, London: IEE Conference Publication no. 155.

Harger, R. O., 1970, *Synthetic Aperture Radar Systems Theory and Design*, New York: Academic Press.

Howland, Paul E., *Television Based Bistatic Radar,* thesis, School of Electronic and Electrical Engineering, University of Birmingham, England, September 1997. Related information is also in P. E. Howland, "A Passive Metric Radar Using a Transmitter of Opportunity," *Proc. International Conference on Radar 1994,* pp. 251–256, May 1994; H. D. Griffiths and N. R. W. Long, "Television Based Bistatic Radar," *IEE Proc.,* Part F, Vol. 133, pp. 649–657, December 1986; H. D. Griffiths, et al., "Bistatic Radar Using Satellite-Borne Illuminators of Opportunity," *Proc.IEE International Conference,* Radar 92, pp. 276–279, 1992; D. Poullin and M. Lesturgie, "Radar Multistatic a Emissions non Cooperatives," *Proc. International Conference on Radar 1994,* Paris, pp. 370–375, May 1994; and B. Carrara, et al., "Le Radar MUET (Radar Multistatique Utilisant des Emetteurs de Television)," *Proc. International Conference of Radar 1994,* Paris, pp. 426–431, May 1994.

http://rcs.ee.washington.edu/rrsl/Projects/Manastash/.

Jelalian, A., 1977, "Laser Radar Theory and Technology," in E. Brookner (Ed.), *Radar Technology,* Norwood, MA: Artech House.

Kildal, P., T. Petterson, E. Lier, and J. Aas, 1988, "Reflectors and Feeds in Norway," *IEEE Antennas & Propagation Society Newsletter,* Vol. 30, No. 2, April 1988.

Kovaly, J. J., 1976, *Synthetic Aperture Radar,* Norwood, MA: Artech House. This is a collection of 33 reports and papers providing in-depth, highly technical treatment of virtually all aspects of synthetic aperture radar.

Rogers, E. E., and R. P. Ingalls, 1970, "Radar Mapping of Venus with Interferometric Resolution of the Range-Doppler Ambiguity," *Radio Science,* No. 2, pp. 425–433, February 1970.

Schleher, D. Curtis, 1991, *MTI and Pulsed Doppler Radar,* Norwood, MA: Artech House.

Siegel, K. M., J. W. Crispin, and R. J. Newman, 1968, "RCS Calculation of Simple Shapes—Bistatic," in J. W. Crispin and K. M. Siegel (Eds.), *Methods of Radar Cross-Section Analysis,* New York: Academic Press.

Skolnik, M. I., 1970, *Radar Handbook,* New York: McGraw-Hill.

Skolnik, Merrill I., 1980, *Introduction To Radar Systems,* 2nd ed., New York: McGraw-Hill.

Skolnik, Merrill I., 2001, *Introduction To Radar Systems,* 3rd ed., New York: McGraw-Hill.

Stimson, George W., 1998, *Introduction To Airborne Radar,* 2nd ed., Raleigh, NC: SciTech Publishing, Part VII, High Resolution Ground Mapping and Imaging, Chapters 31–33.

Weiner, S., 1984, in Carter, A. B., and D. N. Schwartz (Eds.), *Ballistic Missile Defense,* p. 94, Washington, DC: Brookings Institution.

8

Loose Ends of Radar Lore

HIGHLIGHTS

- Picking up on loose ends of radar lore
- The radar line of sight
- The ionosphere as absorber, refractor, reflector, and polarizer
- The troposphere, complex but definable
- Deriving the far field of an antenna, and a target
- Some radar rules of thumb

Several other brief quantitative notions and a list of rules of thumb should be included in any set of radar fundamentals. Some are good for general day-to-day use; others are important to detailed radar design calculations. They are pulled together in this chapter without concern for their relevance to each other and only because they are valid radar lore.

8.1 RADAR LINE OF SIGHT

Solving for the optical horizon at a given range is a simple problem in trigonometry, as shown in Figure 8.1. The range to the horizon is given in Equation (8.1). Radio frequency waves refract (bend) as they propagate through the many layers of the atmosphere [Meeks, 1982]. To account for refraction, we multiply the Earth's radius by a refraction factor, as given in Equation (8.2). Refraction factors that are constant as a function of altitude are often used. Common constant refraction factor values are 4/3 and 6/5. Refrac-

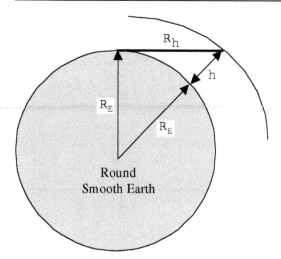

Figure 8.1 Range to the horizon.

tion factors that are exponential with altitude are also used. Often, an approximation to the range to the horizon is used because $2k_r R_E h \gg h^2$. This approximation is acceptable for the vast majority of ground-based and airborne radar geometries. This approximation is *unacceptable* for computing the associated elevation angle. Thus, it is best to just use the actual range to the horizon equation.

$$R_h^2 + R_E^2 = (R_E + h)^2$$

$$R_h = \sqrt{2R_E h + h^2}$$

(8.1)

$$R_h = \sqrt{2k_r R_E h + h^2}$$

(8.2)

where:

R_h = range to the horizon, meters
R_E = radius of the Earth, 6371 kilometers
h = radar or target height, meters
k_r = refraction factor, no units

Two range-to-the-horizon components make up the radar line of sight (LOS), the clutter horizon, and the target horizon, as shown in Figure 8.2. The clutter horizon is how far the radar can illuminate the Earth's surface, thus producing a clutter return [Equation (8.3)]. The target horizon is how far a target is when it comes over the horizon [Equation (8.4)]. The radar LOS, or radar horizon, is the sum of the clutter horizon and the target horizon, as given in Equation (8.5).

$$R_{hc} = \sqrt{2 k_r R_E h_r + h_r^2} \tag{8.3}$$

$$R_{ht} = \sqrt{2 k_r R_E h_t + h_t^2} \tag{8.4}$$

$$R_{LOS} = R_{hc} + R_{ht} \tag{8.5}$$

where:
R_{hc} = clutter horizon, meters
h_r = radar height, meters

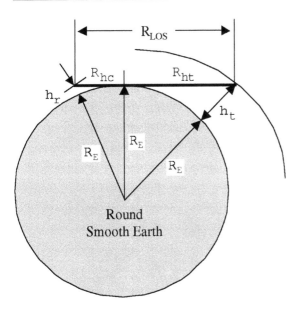

Figure 8.2 The radar line of sight.

R_{ht} = target horizon, meters
h_t = target height, meters
R_{LOS} = radar line of sight, meters

8.2   Properties of the Propagation Medium

In all discussions in this book, we have considered the propagation medium lossless and noninterfering. Actually, it is not. In fact, both the ionosphere (for low microwave frequencies) and the troposphere (for high microwave frequencies) cause attenuation, refraction, and dispersion of radar signals. Because the degradation caused by dispersion is usually small compared to the first two and relatively complex, we shall not discuss it. Tropospheric refraction was discussed in the previous radar LOS section.

8.3   Ionosphere

Electromagnetic waves are attenuated, refracted, and reflected, and their polarization is rotated by the ionosphere [Glasstone and Dolan, 1977]. Ampere's and Faraday's laws allow us to understand the ionosphere in principle, but its condition of constant fluctuations due to energy deposited continuously by the sun, the cosmos, and man make it impossible to predict accurately.

If we think of the ionosphere as a large, imperfectly conducting sheet with conductivity a variable depending on the density of free electrons in the medium, we have a model that will serve our purposes. When there are moderate numbers of free electrons, there is refraction. When these electrons collide with other atoms, there is attenuation; when the electron density is very high, there is reflection. All these interactions are more pronounced at longer wavelengths; in fact, there is a λ^2 dependency.

8.3.1   Attenuation

Attenuation of electromagnetic waves by interaction with an ionized medium is given in Equation (8.6). Because the altitude of the ionosphere exceeds 45 Nmi where $v \leq 6 \times 10^6$, $v^2 \ll \omega^2$ for radars of interest and can be dropped out, yielding Equation (8.7). When the path is oblique through some path length, appropriate trigonometric functions must be applied. For the two equations on attenuation to be useful, we must know ionospheric electron densities at various levels. Unfortunately, these vary enormously from day to night with solar activity and as a function of latitude. In practice, radars that must operate through the ionosphere take frequent soundings as well as subscribe to National Bureau of Standards periodicals on the matter.

$$\alpha = \frac{2.6 \times 10^4 N_e v}{\omega^2 + v^2} \tag{8.6}$$

$$\alpha \cong 4 \times 10^{-3} \frac{N_e}{f^2} \tag{8.7}$$

where:

α = attenuation, dB/Nmi
N_e = electron density, electrons/cm^3
v = collision frequency of the electrons with particles
ω = radar frequency, radians/second
f = radar frequency, megahertz

8.3.2 Refraction

Useful refraction relationships for the ionosphere can also be written down. Loss in an ionized medium has an imaginary component that produces changes in phase that in turn result in an effective change in direction of the radar beam. An index of refraction based on electron densities is given in Equation (8.8). This equation presumes radar frequencies above 10 MHz and N_e below 10^9 per cm^3. Outside these bounds, the plasma frequency ($f_p \approx 9\sqrt{N_e}$) and the collision frequency modify the relationship substantially. From the geometry of refraction shown in Figure 8.3, calculations for refraction are easy when the electron densities of ionospheric layers are known.

$$\eta \cong \left(1 - \frac{N_e}{10^4 f^2}\right)^2 \tag{8.8}$$

where:

η = index of refraction

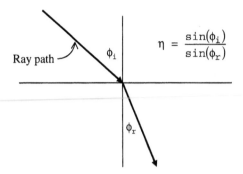

$$\eta = \frac{\sin(\phi_i)}{\sin(\phi_r)}$$

Figure 8.3 Refraction geometry.

8.3.3 Polarization Rotation

Faraday rotation is another phenomenon that affects linearly polarized electro-magnetic waves propagating through an ionized medium existing in a magnetic field. Virtually all the arriving energy interacts with the ionized particles. Conceptually, electrons excited by the incident electric field of the radar wave are diverted in their paths by the presence of the Earth's magnetic field in accordance with Faraday's law. Thus, they oscillate and reradiate at an angle from the direction of their excitation. With each stage of excitation, oscillation, and reradiation, the angle increases. The result is that the polarization (E-field) of the radiated energy revolves. As with the other ionospheric interactions discussed, Faraday rotation has a λ^2 dependence. The actual rotation experienced depends on the path length, the Earth's magnetic field strength, and the angle between the magnetic field lines and the direction of propagation. A worst case (100-Nmi path, zero elevation angle, maximum interaction angle, high field strength, and electron density) relationship has been worked out as given in Equation (8.9). It is apparent there is a substantial rotation effect (>10°) at frequencies as high as 2.5 GHz.

$$\Omega = \frac{6.3 \times 10^{20}}{f^2} \tag{8.9}$$

where:

Ω = Faraday rotation, degrees

8.4 TROPOSPHERE

Tropospheric or atmospheric attenuation is the loss in power in the radar wave as it travels through the molecules in the atmosphere [Morchian, 1990]. At radar frequencies, oxygen and water vapor are the offending molecules. Atmospheric attenuation is a result of an electromagnetic resonance between the radar wave and oxygen and water vapor molecules. Thus, atmospheric attenuation is a function of wavelength. As shown in Figure 8.4, the lower the radar frequency, the lower the attenuation. Because the distribution of oxygen and water vapor varies with altitude, atmospheric attenuation varies with altitude. As shown in Figure 8.5, the higher the altitude, the lower the attenuation. Rain and clouds also effect atmospheric attenuation. It is best presented in Figure 8.6, which is based on data and calculations reported by Barton (1979, pp. 468–472) and Nathanson (1969, pp. 193–199). Included in Figure 8.6 are absorption effects of various atmospheric molecules, especially oxygen and water vapor, as well as rain and clouds.

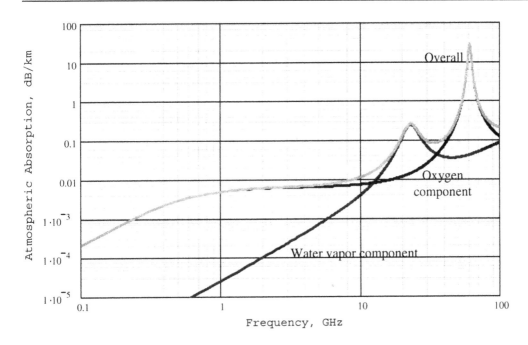

Figure 8.4 Atmospheric attenuation—overall with oxygen and water vapor components.

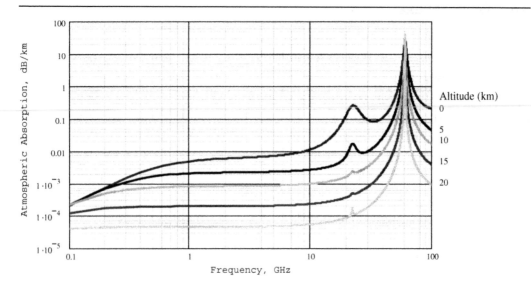

Figure 8.5 Atmospheric attenuation as a function of altitude.

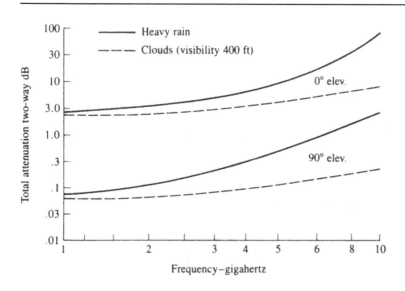

Figure 8.6 Atmospheric attenuation (dB/mile)—rain and clouds.

8.5 FAR FIELD OF AN ANTENNA, AND A TARGET

The far field of an antenna can be defined as the distance from the antenna beyond which there is "negligible" phase difference between wavelets arriving
from the center of the aperture and wavelets arriving from its edge. The aperture is assumed to be "focused" on infinity, and the far field begins when the radiating energy is essentially a plane wave. The far-field relation is easy to derive
by referring to Figure 8.7, as given in Equation (8.11). What value should we assign arbitrary phase difference? Conventional wisdom has it that, when at least
half the power arriving from the antenna edge is contributing to the pattern,
the far field has been reached. This occurs at a phase difference of 45° ($\lambda/8$).
The far-field distance for a $\lambda/8$ phase difference is given in Equation (8.11). A
more conservative engineer might define the phase difference as $\lambda/16$, a less
conservative one as $\lambda/4$. Their "far fields" would vary accordingly.

$$(R+\varepsilon)^2 = R^2 + \left(\frac{D}{2}\right)^2$$

$$R \cong \frac{D^2}{8\varepsilon} \tag{8.10}$$

$$R_{ff} \cong \frac{D^2}{\lambda} \tag{8.11}$$

where:
  R = far field distance, meters
  ε = arbitrary phase difference from center to edge of the antenna, meters

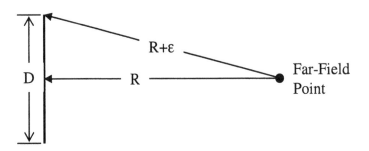

Figure 8.7 Far-field calculation.

D = antenna dimension, meters
R_{ff} = far field distance, meters
λ = wavelength, meters

There is also a far field associated with target RCS. This is because most complex radar targets are made of multiple scatterers. Each individual scatterer produces a complex (amplitude and phase) reflection, and these combine in space as the reflected wave propagates back to the radar. Defining the far-field distance for target RCS is more challenging than for an antenna. This is because it is difficult to know which scatterers are combining to produce the overall RCS. This information can be gained from detailed RCS measurements and/or computer modeling.

8.6 CONVENIENT RADAR RULES OF THUMB

Some results of radar design problems occur repeatedly. They seem to represent enduring verities. They are recognized as having broad application and crop up in meetings or are used as springboards for quick extrapolations. A radar neophyte needs to memorize a handful of such rules of thumb as are presented here.

- There are approximately 7 μsec to a kilometer (two-way).

- There are approximately 11 μsec to a statute mile (two-way).

- There are 12 μsec to a nautical mile (two-way).

- One second is 150 m or approximately 500 ft (two-way).

- A 1-MHz bandwidth gives 150-m range resolution.

- A 500-MHz bandwidth gives approximately 1-ft range resolution.

- The natural limit on range resolution is one RF cycle.

- There are 4π steradians or approximately 41,000 square degrees in a sphere.

- One hertz of Doppler shift is 1 m/sec of range rate at VHF (λ = 1 m); 20 Hz of Doppler is 1 m/sec of range rate at C-band (λ = 0.1 m); and approximately 60 Hz of Doppler shift is 1 m/sec of range rate at X-band (λ = 0.033 m).

- One hertz of Doppler shift is 1 ft/sec of range rate at UHF (λ = 2 ft); 24 Hz of Doppler is 1 ft/sec of range rate at X-band (λ = 1 in).

- The natural limit on angular resolution is λ^2.

- One-tenth milliradian angular accuracies are all the atmosphere will support; ditto 0.1-ft range accuracies.

- A 300-K receiver has approximately a 3-dB noise figure.

- Probability of detection is meaningless without probability of false alarm.

- The gain all the way around any antenna is 0 dB.

- A phased array of n isotropic elements has n beam positions, sidelobes of $1/n$, and gain of n.

- Radar maps as good as optical photographs are impossible.

- A fixed antenna, focused on infinity, has a maximum (finest) cross-range resolution equal to its diameter.

- A synthetic aperture antenna can have a maximum (finest) cross-range resolution equal to half its physical aperture, no matter how small.

- A passive target cannot reradiate more energy than it intercepts.

- A bistatic radar cross section of a smooth body is equal to the monostatic radar cross section at one-half the bistatic angle [Siegel, Crispin, and Newman, 1968, p. 158].

8.7 EXERCISES

1. Using electron densities of 10^4 and $10^5/\text{cm}^3$, calculate some ionospheric effects (attenuation, refraction, and polarization rotation) for a microwave frequency $f_c = 3$ GHz. Then introduce tropospheric effects.

2. What is the far field of the Arecibo antenna on its boresight?

3. Fifteen ground-based radars stretched across the northern border of the United States will give coverage at 1400 m altitude and above. How many ground-based radars will be required to extend coverage down to 60 m? Use a refraction factor, $k_r = 4/3$.

4. How many airborne pulse-Doppler radars flying at 6000 m are required to cover the United States northern border against targets flying down to 30 m? Use a refraction factor, $k_r = 4/3$.

5. The space track radar in Florida has an antenna aperture diameter of 55 m at UHF ($\lambda = 0.6$ m). What is the range to its far field?

6. A homemade telescope has an aperture of 0.2 m. What is the range to its far field? (Take optical wavelengths to be 0.5 µm.)

7. What is the radar line of sight for a radar height h_r = 10 m and a target height h_t = of 2000 m? Use a refraction factor, $k_r = 4/3$.

8.8 REFERENCES

Barton, D. K., 1979, *Radar Systems Analysis*, Norwood, MA: Artech House.

Glasstone, S. and P. J. Dolan (Eds.), 1977, *The Effects of Nuclear Weapons*, U.S. Department of Defense and Energy, R&D Admin., Washington, D.C. The presentation on the ionosphere is based an the excellent tutorial on the subject contained on pages 462–466 and 489–494.

Meeks, M. L., 1982, *Radar Propagation at Low Altitudes*, Norwood, MA: Artech House. This is a classic reference on radar wave propagation, Meeks is concise and to the point.

Morchin, William C., 1990, *Airborne Early Warning Radar*, Norwood, MA: Artech House. Chapter 3, pp. 89–98, is the source of the algorithm used to generate the atmospheric attenuation figures.

Nathanson, F. E., 1969, *Radar Design Principles*, New York: McGraw-Hill.

Siegel, K. M., J. W. Crispin, and R. J. Newman, 1968, "RCS Calculation of Simple Shapes—Bistatic," in J. W. Crispin and K. M. Siegel (Eds.), *Methods of Radar Cross-Section Analysis*, New York: Academic Press.

9

Radar Potentials and Limitations

From the previous chapters, you can adequately understand and summarize what radars do and how they do it. However, you need some perspective on whether radar is the system of choice for accomplishing a particular objective in view of some of the alternatives available. Good managers must be able not only to understand the salient technical verities of the work they are managing, but be able to make a judicious choice among the options available for getting the job done.

Allocation of scarce resources is probably a manager's most vital task. Reviewing the key attributes and limitations of radar and knowing how other technologies and other systems compare in capability will assist you greatly. Doing these comparisons with matrices and lists is not as much fun or as illuminating as doing them by missions. The first approach would be similar to preparing a cookbook to satisfy all tastes; the second is to discuss the menu that will satisfy a particular craving. Some of these missions are big, and some are small. The extent of the analytical discussion of them has no relationship to their size, but to the adequacy of radar for meeting the requirement.

9.1 SURVEILLANCE

If the range is neither too long nor too short (less than several thousand kilometers in the first case and a kilometer or so in the second), radar is preeminent in any surveillance mission that requires night and bad-weather operations. This is a very broad mission. It includes vehicle traffic on the highways and at state or national borders, aircraft on runways and in the air, and ships at sea and in harbors. In most cases, it not only includes surveillance of the vehicles, but radars aboard the vehicles themselves (for example, altimeters). Boats maneuvering in harbors in the United States do not usually receive centralized direction; they need onboard radars. Aircraft in airports receive centralized direction; they generally do not have surveillance radars. The mission does not include underwater surveillance. Electromagnetic waves do not propagate well underwater; sound waves do. Sonar is the underwater analog of radar and is used there.

Radar is preeminent in this mission for two reasons: It is active, and it is usually all weather. Being active allows night operations; the radar provides its own illumination. A radar's measuring stick is the speed of light. Nothing in the universe is faster or more efficient. Even if measurements of position could be made by other means (and they can in many cases), the radar measurement is simpler and easier. Imagine a sonar for traffic patrol: bigger, heavier, requiring more power (the medium is lossy), and with a speed of propagation of 330 m/sec rather than 3×10^8.

Wavelength determines a radar's all-weather capability. The longer the wavelength, the less problem there is with weather. Although radar is thought of as all weather, not all radar is all weather. Radars in the infrared or the visible light spectrum (lasers) are not all weather. Radars in the microwave region and below are all weather.

As ranges get longer or the need for better resolution emerges, radar begins to falter. An example of the first condition is deep space surveillance; an example of the second is perimeter security of a plant, base, or other facility.

There are particular radars that can find satellites at geosynchronous altitude (40,766 km above the equator) if they know in general where to look. (The Haystack radar of the Lincoln Laboratory is one. It has a 36.6-m reflector antenna at X-band, a theoretical 70-dB gain, and average power in the hundreds of kilowatts.) However, the task of surveying space for satellites at ranges of more than a few thousand kilometers is too much for radar to undertake. The extent of the problem for space surveillance is not difficult to adumbrate, because geosynchronous satellite orbits are deterministic. By the laws of celestial mechanics, a satellite must pass across the equator twice in each orbit. The Earth turns under each orbit at the rate of 15° per hour. The satellite's inclination angle is the highest latitude over the Earth to which its orbit carries it. A combination of those factors determines at what times, in what directions, and at what ranges satellites in circular orbits will pass over a region on the Earth's surface. Satellites in elliptical orbits add to the difficulty, because the only constraint on the duration of their orbits is that of remaining an Earth satellite.

With them, there is no association of altitude with velocity, although a line from the satellite to the Earth's center will carve out equal areas in equal time.

As a little calculation using the radar range equation of Chapter 1 will show, radars of the kind we have described (power-aperture products of 3×10^9 kW-m^2) do not have time to detect 1-m^2 RCS at 6400 km and scan a substantial segment of the sky as well. Optical telescopes placed at correct locations on the Earth's surface can do well at accounting for these longer-range satellites—not with the same rapidity of response, because they cannot operate in daylight, but with better overall accuracy.

At 6400 km, the segment of the spherical shell of possible satellite orbits eclipsed by the Earth is 1/16 the area of the entire shell, so these satellites will be illuminated by the sun 15/16 of the time. For higher altitude satellites, the eclipsing is less. Therefore, using a passive optical satellite for high-altitude satellite surveillance may be a good idea.

Radar also tends to be less than ideal in a surveillance role when very high resolution is required. For example, a sensor covering the outside perimeter of a security fence would need to differentiate among vehicles, animals, and human beings. How many resolution cells are required for recognition of various classes of target is a matter of intense debate, but it is clear that the resolution (perhaps 1 m) of a very narrow beam, millimeter wave radar at a few hundred yards could not compete with an optical system of the same diameter (which would have several orders of magnitude better angular resolution) or even the human eye (which can resolve at least 0.3 m at 450 m in nonideal conditions).

At these comparatively short ranges, weather and night conditions do not affect passive systems appreciably, so infrared sensors and low-light-level TV are very competitive, even when there is considerable dust, fog, and smoke.

9.2 NAVIGATION

The Earth is literally webbed with long- and short-range navigation systems that use electromagnetic waves but not radar. In many of these, the vehicle is passive, the system itself providing virtually all the energy required. In others, the vehicles have transponders. In addition, the Navstar Global Positioning System (GPS), a net of satellites, provides positive location to anyone in the world to a three-dimensional accuracy of a few meters or better.

Yet, radar has a navigation mission. It can provide essential Doppler information to update an inertial guidance system, and radar maps can provide references by which a vehicle can locate itself. Because commercially available inertial navigators have 0.45-m/sec or so drifts, they need updating. A Doppler radar is one solution. There are others, GPS being the primary example. There are also other means for obtaining map references: optical, IR, and microwave radiometry. If the other forms of navigation are good enough, maps are not required.

Both Doppler radars and radar mapping are essential in military operations. In such an emergency, it is necessary to be self-contained independent of time of day, weather, and the status of outside navigation aids. In that situation, radar shines. Another form of navigation is weapon guidance. Radar has a role in one form of this: active and semiactive homing. Because we can guide these vehicles in several other ways, radar's role could be described as important but not unique.

9.3 SIGNATURES

We have already emphasized that electromagnetic wavelengths that allow radar to penetrate fog, dust, smoke, and rain are effective because they are long compared to infrared and light waves. These same waves are less effective as imagers because their resolution is not fine enough. The physical limit of angle resolution is about λ^2 and, for range resolution, a few RF cycles. If we want details to use in recognition, identification, analysis, targeting, and so forth, there is no substitute for resolution. As has been mentioned in other chapters, the wavelengths at microwave frequencies (C-band) are 5×10^{-2} m. Wavelengths of visible light are 5.5×10^{-7} m, of far infrared, 20×10^{-6} m. A radar image has, say, 10 by 10 resolution cells. The same image in the visible has 1,000,000 by 1,000,000. In the far infrared, it has 25,000 by 25,000.

Of course, we have radars at optical frequencies, and they make high-resolution images (which are not as appealing to the human eye and do not contain as much information as images produced by diffused illumination) in three dimensions. Photography and television are just as convenient and far less expensive. Moreover, infrared images can be obtained at night, using the irradiance of the objects in the scene as the illumination source.

Radar imagery must resort for its utility to the all-weather role. Again, the military applications dominate, although natural disasters have substantial relevance. The military commander needs to know the status of an enemy's deployed forces and cannot wait for daylight or good weather; attacking aircraft must be able to identify enemy targets through snow, rain, fog, smoke, and dust. In these situations, some resolution is urgent, but super-resolution is a luxury. Here, modern radars (all that the technology can provide) are indispensable.

There is a partial way out of the dilemma of all-weather imagery. It is to place the image-gathering sensor so close to its subject that the weather no longer interferes. Although this approach is no help in obtaining high-resolution maps of a battlefront or a hurricane's passage, it is promising for weapon delivery where the weapon can be placed near the target by various means and can then use onboard high-resolution sensors to close with it autonomously. Whether such applications are affordable remains to be seen.

9.4 SCIENCE

Speculation on the potential uses of radar in science is for the futurists. Current and past uses are substantial, however. The study of the ionosphere and of our own and neighboring planets are cases in point. Study of the ionosphere, and particularly of the auroral region in the northern hemisphere, has proceeded vigorously for about 30 years. Much knowledge about its characteristics, particularly propagation, has been exploited in system design.

Mapping radars have been flown several times in space by NASA. Because electromagnetic waves have some penetrating power, images of subsurface features in both land and water were obtained, portending new dimensions in Earth resources research.

The Haystack radar was used in the 1960s in a test of relativity by making measurements on the planet Mercury as the radar line of sight to Mercury neared the edge of the sun. The Arecibo system has examined all the planets inside Jupiter. Plans have been made to study the planets further with a new S-band radar system. Planets such as Venus (shrouded in an atmosphere that is impenetrable to many of the higher frequencies) are made to order for radars. We have already seen high-resolution synthetic aperture radar maps of the surface of Venus. Venus is being explored from Earthbound radars and from radars on board the space vehicles sent there. In fact, virtually all explorers of the solar system that expect to land will have at least a radar altimeter on board, barometric altimeters being useless when there is no atmosphere and unsatisfactory when the atmosphere pressure gradients are poorly known.

Although radars have not achieved the eminence of ultrasonic, infrared, and X-ray medical diagnostic techniques, there are conceptual uses of parts of the spectrum for imaging internal tissue and organs. However, the frequencies that penetrate best on the macro scale are too low for imaging, and the higher frequencies, when stopped by tissue, deposit their energy (as with microwave ovens and diathermy). The ultrasound systems used to image the heart and arteries near the heart are "radars of sound" (sonar). X-ray radars have long been discussed, the problems of focusing rays with the energy of X-rays and the problem of controlling the backscattered energy (it does not return as X-rays) being the principal barriers to a useful system. X-rays have high penetration on the micro scale, but their interactions with tissue are atomic rather than electromagnetic.

9.5 SUMMARY

To summarize, the qualities that make radar essential to modern life also limit its utility. The fact that it furnishes its own illumination makes it valuable for obtaining almost instantaneous range and range rate data on targets and for night operations against passive or noncooperative targets, but this limits its ultimate range. The fact that the frequencies at which radars operate can penetrate bad weather, smoke, and dust and can see over the Earth's horizon makes them essential to emergency operations, particularly military operations, but prevents them from obtaining the high resolution they need to get fine-grained images.

The vigorous application of new technology and human ingenuity will tend to expand radar's applications and mitigate its shortcomings. However, the natural laws cannot be overcome, inevitably consigning radar to its niche: an interesting and important technology with multifaceted but limited uses.

Conversion to Decibels

Bels were named after Alexander Graham Bell. They were simply the logarithm of the ratio of power out and power in. The bel was too small to be useful, so the decibel quickly emerged. The definition of a decibel (dB) is contained in any radio engineer's handbook, as given below.

$$dB = 20 \log_{10}\left(\frac{V_o}{V_i}\right) = 10 \log_{10}\left(\frac{P_o}{P_i}\right)$$

where:
  dB = decibel, no units
  V_o = voltage out, volts
  V_i = voltage in, volts
  P_o = power out, watts
  P_i = power in, watts

In colloquial usage, decibels are used to express any ratio or often just an absolute value. Decibels are just ten times the $\log_{10}$ of that ratio or value. A factor of two results in about 3 dB, while a factor of one-half results in about –3 dB.

$$dB = 10 \log_{10}(\text{absolute}) \qquad \text{absolute} = 10^{\left(\frac{dB}{10}\right)}$$

$$3.0103 = 10 \log_{10}(2) \qquad -3.0103 = 10 \log_{10}(0.5)$$

Because radar engineers often deal in power, decibels relative to a watt or decibels relative to a milliwatt are used. Another common use of decibels is for

RCS, decibels relative to a square meter. Some approximate decibel values usually memorized by radar people are shown in Table A1.1.

$$dBW = 10 \log_{10}\left(\frac{P(W)}{1 \text{ watt}}\right) \qquad dBm = 10 \log_{10}\left(\frac{P(mW)}{1 \text{ milliwatt}}\right)$$

$$dBW = dBm - 30 \qquad dBm = dBm + 30$$

$$dBsm = 10 \log_{10}\left(\frac{\sigma(m^2)}{1 \text{ m}^2}\right)$$

where:

dBW = decibels relative to one watt, carries the units of watts

$P(W)$ = power, watts

dBM = decibels relative to one milliwatt, carries the units of milliwatts

$P(mW)$ = power, milliwatts

dBsm = decibels relative to one square meter, carries the units of square meters

$\sigma(m^2)$ = radar cross section, square meters (m^2)

Table A1.1

Absolute	dB	Absolute	dB
$0.000001 = 10^{-6}$	−60	$1,000,000 = 10^6$	60
$0.00001 = 10^{-5}$	−50	$100,000 = 10^5$	50
$0.0001 = 10^{-4}$	−40	$10,000 = 10^4$	40
$0.001 = 10^{-3}$	−30	$1,000 = 10^3$	30
$0.01 = 10^{-2}$	−20	$100 = 10^2$	20
$0.1 = 10^{-1}$	−10	$10 = 10^1$	10
$1 = 10^0$	0	$1 = 10^0$	0

The Radar Spectrum

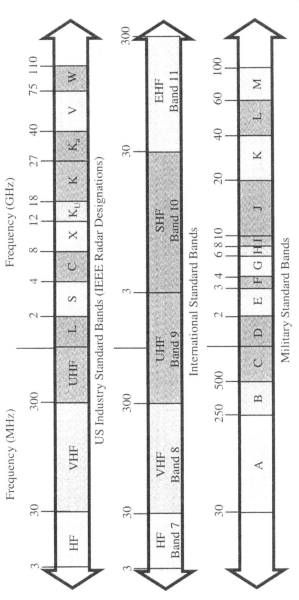

Appendix 3

Fourier Series and Transforms

Working in the field of heat transfer in the early nineteenth century, Jean B. S. Fourier found that virtually all functions of time, particularly repetitive ones, could be described in a series of sine and cosine waves of various frequencies and amplitudes. His work has been described as one of the most elegant developments in modern mathematics. Whatever its stature for the world, the benefits for the radar engineer are epic. The following presentation uses the approach taken by H. H. Skilling in *Electrical Engineering Circuits*, Chapters 14 through 15 (New York, Wiley, 1957). George Stimson also has an excellent discussion on Fourier series and transforms in *Introduction To Airborne Radar*, 2nd ed., Chapters 17 and 20 (Raleigh, NC: SciTech Publishing, 1998).

A3.1 FOURIER SERIES

The statement for the Fourier series is that any wave may be broken down into the sum of sines and cosines of various amplitudes and frequencies. In mathematical notation,

$$f(t) = \frac{1}{2}a_0 + a_1 \cos(\omega t) + a_2 \cos(2\omega t) + \dots + b_1 \sin(\omega t) + b_2 \sin(2\omega t) + \dots \quad (A3.1)$$

The term $f(t)$ is a function of time here (it does not have to be), and the a_i and b_i terms are constants that are to be found so as to make the expression on the right equal to $f(t)$; $1/2\, a_0$ is the dc component of the function, if any.

The next step is to find a way to evaluate the a_i and the b_i. To do this, we use the orthogonality of the sine and cosine function, which is that the value of their cross products integrated from zero to 2π is zero. Furthermore, the sine and cosine each integrated from zero to 2π is zero. These integrations are easy to check by referring to a table of integrals; they are

$$\int_0^{2\pi} \sin(mx)\,dx = 0 \quad \int_0^{2\pi} \cos(nx)\,dx = 0 \quad \int_0^{2\pi} \sin(mx)\cos(nx)\,dx = 0$$

(A3.2)

Also,

$$\int_0^{2\pi} \sin(mx)\sin(nx)\,dx = 0 \quad \int_0^{2\pi} \cos(mx)\cos(nx)\,dx = 0$$

(A3.3)

But,

$$\int_0^{2\pi} (\sin(mx))^2\,dx = \pi \quad \int_0^{2\pi} (\cos(nx))^2\,dx = \pi$$

(A3.4)

Let us pick a coefficient to solve for, say, a_2. Multiply through Equation (A3.1) by $\cos(2\omega t)$, then multiply both sides by $d(\omega t)$ and integrate. We get

$$\int_0^{2\pi} f(t)\cos(2\omega t)\,d(\omega t) = \int_0^{2\pi} \frac{1}{2}a_0\cos(2\omega t)\,d(\omega t) +$$

$$\int_0^{2\pi} a_1\cos(\omega t)\cos(2\omega t)\,d(\omega t) + \ldots \int_0^{2\pi} b_1\sin(\omega t)\cos(2\omega t)\,d(\omega t) + \ldots$$

We can see by inspection that most of these integrals are zero. On the right side of the equation, the first and second terms are zero. However, the third term equals $a_2\pi$. In fact, all other terms written down and all implicit terms are zero, giving finally

$$\int_0^{2\pi} f(t)\cos(2\omega t)\,d(\omega t) = a_2\pi$$

(A3.5)

and

$$a_2 = \frac{1}{\pi} \int_0^{2\pi} f(t) \cos(2\omega t) d(\omega t) \tag{A3.6}$$

Because we do know $f(t)$, we can find a_2, even if we have to do it by numerical integration on a computer or calculator. In addition, because we did this in general, we can now generalize to

$$a_n = \frac{1}{\pi} \int_0^{2\pi} f(t) \cos(n\omega t) d(\omega t) \tag{A3.7}$$

and, after multiplying through by a sine function

$$b_n = \frac{1}{\pi} \int_0^{2\pi} f(t) \sin(n\omega t) d(\omega t) \tag{A3.8}$$

Equation (A3.1) is the Fourier series, but it can be rewritten for brevity as

$$f(t) = \frac{1}{2}a_0 + \sum_{m=1}^{\infty} (a_m \cos(m\omega t) + b_m \sin(m\omega t)) \tag{A3.9}$$

Writing down a function and evaluating it is the essence of tedium, but by astute observation, the process can be shortened. Some of the waves of interest in radar are not too difficult. For example, a periodic square wave (Figure A3.1) has no even harmonics. It is just an infinite sum of odd harmonics with ever-increasing frequencies and ever-decreasing coefficients. The answer for a square wave is

$$f(\omega) = \frac{4}{\pi}\left((\sin(\omega t)) + \frac{1}{3}\sin(3\omega t) + \frac{1}{5}\sin(5\omega t) + \ldots \right) \tag{A3.10}$$

A3.2 FOURIER TRANSFORMS

To arrive at the Fourier transforms, we must take a few preliminary steps. The first is to restate Equation (A3.1) in exponential form. We can do this by substituting for cosines and sines their identities, which are

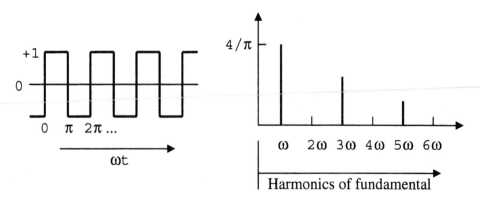

Figure 12.1 A square wave and its Fourier spectrum.

$$\cos(x) = \frac{e^{jx} + e^{-jx}}{2} \qquad \sin(x) = \frac{e^{jx} - e^{-jx}}{2j}$$

We can now write the Fourier series as

$$f(t) = \ldots + A_{-2}e^{-j2\omega t} + A_{-1}e^{-j\omega t} + A_0 + A_1 e^{j2\omega t} + A_2 e^{j2\omega t} + \ldots \qquad \text{(A3.11)}$$

or

$$f(t) = \sum_m A_m e^{j\omega t}$$

The coefficients are related in this way:

$$A_n = \frac{1}{2}(a_n - jb_n) \qquad A_{-n} = \frac{1}{2}(a_n + jb_n) \qquad \text{(A3.12)}$$

and the coefficients can be found from the integral

$$A_n = \frac{1}{2\pi}\int_0^{2\pi} f(t)e^{-jn\omega t}\,d(\omega t) \qquad \text{(A3.13)}$$

The term $f(t)$ is the function of time to be expressed as a Fourier series, and the integer n can now be positive, negative, or zero. The integral (13) is used just like Equations (A3.7) and (A3.8) to obtain coefficients. It is more efficient, because there is only a single integral, and it is much more readily integrated. Equation (A.13) can be derived from Equations (A3.7) and (A3.8) using Euler's theorem. By matching their infinite series, Euler showed that

$$e^{j\alpha} = \cos(\alpha) + j\sin(\alpha)$$

Consider now a square wave, as in Figure A3.2, where we have made the width of the pulse equal the interpulse period divided by a constant. We solve for the coefficients with

$$A_n = \frac{1}{2\pi} \int_{-\pi/k}^{\pi/k} e^{-jnk} dx$$

for $n = 0$, $A_0 = 1/k$, and for $n \neq 0$, $A_n = \dfrac{1}{k}\dfrac{\sin(n\pi/k)}{n\pi/k}$

giving a spectrum like that of Figure A3.3. The spectrum of Figure A3.3 has an envelope of the $\sin(x)/x$ shape, which we know to be the Fourier transform of a single pulse and the larger the value of k, the closer the spectral lines are to-

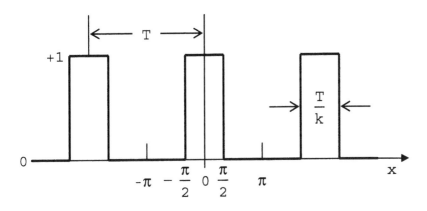

Figure 12.2 Another square wave.

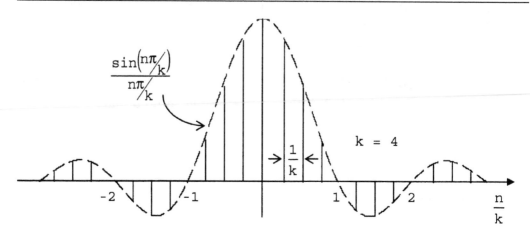

Figure 12.3 Another spectrum of a square wave.

gether and the more closely a continuous function is replicated. No matter how large we make k, however, this envelope is still caused by lines of discrete frequencies. How can those discrete frequencies exist? They exist because the time function they are describing goes on forever. If it starts and stops, there will be, of necessity, a continuum of frequencies; that is, the A_n will blend. There will be an A_n for every incremental distance along the frequency axis, not just at discrete points. The A_n will become a function of frequency, say, $g(\omega)$, and the integer n will become an on the other side of the equation. The limits of integration must be extended to find the stopping point of the time function. These changes give us

$$g(\omega) = \frac{1}{2\pi} \int_{-\infty}^{\infty} f(t)e^{-j\omega t}\,dt \qquad\qquad (A3.14)$$

for the frequency function and, by symmetry

$$f(t) = \int_{-\infty}^{\infty} g(\omega)e^{j\omega t}\,d\omega \qquad\qquad (A3.15)$$

These are called the *Fourier integral equations* or the *Fourier transform pair*. In this book, waveform time functions are transformed into their spectra. For this transform, Equations (A3.14) and (A3.15) are applicable. Let us do an example with a single square pulse of height one (1) and duration τ:

$$g(\omega) = \frac{1}{2\pi} \int_{-\tau/2}^{\tau/2} 1 e^{-j\omega t} dt \tag{A3.16}$$

$$g(\omega) = \frac{1}{j\omega 2\pi} \left[-e^{-j\omega\frac{\tau}{2}} + e^{j\omega\frac{\tau}{2}} \right] \tag{A3.17}$$

$$g(\omega) = \frac{1}{\pi\omega} \sin\left(\omega\frac{\tau}{2}\right) = \frac{\tau}{2\pi} \frac{\sin\left(\omega\frac{\tau}{2}\right)}{\omega\frac{\tau}{2}} \tag{A3.18}$$

The results are shown in Figure A3.4. In addition, $g(\omega)$ can be converted back into $f(t)$ rather easily by using the form $g(\omega)$ takes in Equation (A3.17) to make up the integrand and remembering to integrate with $d\omega$ from $-\infty$ to ∞. Fourier transforms can also be used to transform antenna illumination functions into far-field antenna gain patterns. Note the similarity of Equation (A3.18) with an antenna gain pattern equation.

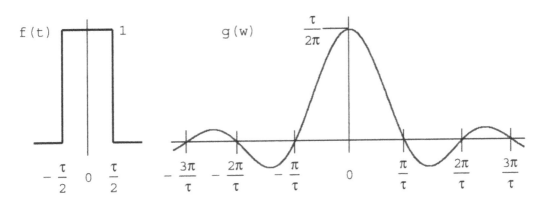

Figure 12.4 A single pulse and its Fourier transform.

Answers to Exercises

The answers listed here are rounded to two decimal places, so do not be concerned if you did not get exactly the same value. A solution set (Mathcad® 11 and RTF files) for the vast majority of the exercises is available from the publisher.

Chapter 1

1. $R = 229.37$ km.

2. Power density $= 6.33 \times 10^{-18}$ W/m^2.

3. $T_0 = 41.496$ K.

4. $C_A = 2 C_P$.

5. $R = 100.5$ km, $\Delta t = 2.56$ sec.

6. $G = 6.28 \times 10^3 \equiv 37.98$ dB.

7. $N = 8.95 \times 10^{-15}$ W $\equiv -140.48$ dBW $\equiv -110.48$ dBm.

Chapter 2

1. IR, $a = 5.64$ mm; RF, $a = 28.21$ m.

2. $N = 400$ elements, $A = 100$ m^2.

3. No answer; exercise is a proof.

4. $G = 6.28 \times 10^3 \equiv 37.98$ dB, $\theta_{3dB\_az} = 1.02°$, $\theta_{3dB\_el} = 2.55°$.

5. Normalized gain at $5° = 4.34 \times 10^{-3} \equiv -23.63$ dB; gain at $5° = 11.37$ dB.

6. $\theta_{3dB} = 4.78°$.

7. $\theta_1 = 14.48°$, wrap-up factor $L/d = 34.64$.

8. $\Delta\theta_0 = 0.196°$.

9. $\theta_{3dB} = 2.76°$ for $25°$, $2.89°$ for $30°$, and $3.54°$ for $45°$.

10. $G = 31.07$ dB for $25°$, 30.88 dB for $30°$, and 29.99 dB for $45°$.

Chapter 3

1. $SNR_d = 13.2$ dB, same for Figure 3.6 and Table 3.1.

2. $n_p = 5$ pulses (rounded up to an integer from 4.6774).

3. $P_d = 0.0036$, 0.0486, 0.2919, and 0.6561 for exactly 1-out-of-4, 2-out-of-4, 3-out-of-4, and 4-out-of-4, respectively; $P_{fa} = 0.2916$, 0.0486, 0.0036, and 0.001 for exactly 1-out-of-4, 2-out-of-4, 3-out-of-4, and 4-out-of-4, respectively; $P_d = 0.9999$, 0.9963, 0.9477, and 0.6561 for at least 1-out-of-4, 2-out-of-4, 3-out-of-4, and 4-out-of-4, respectively; and $P_{fa} = 0.3439$, 0.0523, 0.0037, and 0.0001 for at least 1-out-of-4, 2-out-of-4, 3-out-of-4, and 4-out-of-4, respectively.

4. $P_d = 0.99$.

5. 24, less than 5.

6. $P_d = 0.99$, $P_{fa} = 9.9985 \times 10^{-12}$, and $SNR_d = 13.9$ dB.

7. $P_{dc} = 09533$ and $P_{fa} = 6 \times 10^{-6}$.

8. $R_d = 77.18$ km.

Chapter 4

1. (A) Use Table 4.1 for the RCS of a flat plate. (B) Use the effective area in place of the physical area.

2. $f_c = 22.77$ GHz (average of the two frequencies).

3. $\theta = 0.328°$.

4. No answer; exercise is a proof.

5. $\bar{\sigma} = 1.35 \times 10^3$ m^2.

6. $S/C = 0.0191 \equiv -17.19$ dB.

7. No answer; exercise is a proof.

Chapter 5

1. $\Delta R = 75$ m, $R_u = 500$ km.

2. $\beta = 5$ MHz, $\tau = 1$ msec.

3. Dead zone $= 150$ km.

4. (A) The pulse width needed to provide ΔR_{dot} is much greater than that required for the minimum range. (B) $\tau = 0.2$ µsec, $T = 1$ msec, $PRI =$
   266.67 µsec.

5. (A) $\tau = 0.2$ µsec, $T = 1$ msec, $PRF = 66.67$ kHz. (B) $R_u = 2.25$ km.

6. (A) $D = 17.6$ m. (B) $\tau = 0.1$ µsec, $\beta = 10$ MHz, $T = 0.0167$ sec.

7. (A) $\tau\beta = 20{,}000$. (B) Anti-jam gain $= 20{,}000$. (C) $\Delta R = 1.5$ m. (D) $\Delta f_d = 5$
   kHz.

8. No answer; exercise is a proof.

Chapter 6

1. $J/N = 24.13 \equiv 13.83$ dB, and thus N does not need to be included.

2. $S/J = 1.25 \equiv 0.98$ dB for 1-m^2 RCS; $S/J = 12.52 \equiv 10.98$ dB for 10-m^2 RCS.

3. $R_{bt} = 112.17$ km for 1-m^2 RCS; $R_{bt} = 354.72$ km for 10-m^2 RCS.

4. $ERP_j = 15.92$ kW.

Chapter 7

1. (A) Isolation $= 1.85 \times 10^{-9} \equiv -88.03$ dB. (B) Isolation $= 4.46 \times 10^{-8} \equiv -73.51$
   dB. (C) Direct signal power $= 0.16$ W $\equiv -8.03$ dBW. (D) Direct signal out of
   FM/CW processor $= -81.53$ dBW. E. Needed, the target signal is -51.47 dB
   under the direct signal.

2. (A) $\tau = 0.2$ µsec, $T = 3.33$ msec, $PRF = 18$ kHz. (B) $R_u = 8.33$ km. (C) No, at
   UHF $\Delta R_{dot} = 90$ m/sec and $R_{dotu+} = 5400$ m/sec.

3. For airborne pulse Doppler radars, the clutter spectrum is complex (see Figure 7.13). The rotating helicopter blades and tail rotor and the vibration of the airframe result in a complex target signal at varying Doppler shifts, some lost in the clutter and some separated from the clutter.

4. (A) $f_{d\,max} = \dfrac{2V_{ac}}{d}$ for both ranges. (B) $T_t = \dfrac{R\lambda}{V_{ac}\,d}$ and $\dfrac{2R\lambda}{V_{ac}\,d}$. (C)

$N_f = \dfrac{2R\lambda}{d^2}$ and $\dfrac{4R\lambda}{d^2}$. (D) $\Delta CR = d/2$ for both ranges.

5. Doppler spread is independent of range; transmit time and a number of Doppler filters are proportional to range; however, cross-range resolution is independent of range.

6. 992.9×10^3 cells/sec.

7. $0.6°$ for both focused and unfocused.

8. 1 m.

9. $D = 0.0134$ m.

10. $R_1 R_2 = 844.27$ km^2.

Chapter 8

1. For $N_e = 10^4/\text{cm}^3$: $\alpha = 4.44 \times 10^{-6}$, $\eta = 999.9998 \times 10^{-3}$; for $N_e = 10^5/\text{cm}^3$: $\alpha = 44.44 \times 10^{-6}$, $\eta = 999.9978 \times 10^{-3}$; $\Omega = 7 \times 10^{13}$; atmospheric attenuation in the troposphere $= 7 \times 10^{-3}$ dB/km.

2. $R_{ff} = 133.34$ km.

3. 75 radars.

4. 7 radars.

5. $R_{ff} = 5.04$ km.

6. $R_{ff} = 80$ km.

7. $R_{LOS} = 197.28$ km.

Appendix 5

Glossary

This glossary provides assistance in understanding terms used in this book. There may be more general, more specific, or entirely different meanings for these terms when they are used elsewhere.

Adaptive array: an electronically steerable antenna designed to respond automatically and optimally to a variety of situations

Adaptive receiver: a receiver with circuitry or perhaps programming that enables it to adjust to the characteristics of the incoming signal

AGC: automatic gain control, keeping the excursions of the received signal within bounds with an automatic negative feedback loop

Ambiguous angle: an angle where the characteristics of an antenna pattern are such that there is uncertainty about the angular location of a target (as with an interferometer)

Ambiguous range: the condition that the round-trip time between pulses is insufficient to accommodate all the targets that the radar will see, resulting in uncertainty about target range

Ambiguous range rate: the situation that occurs when the Doppler processing scheme of the radar is such that target range rates fold over, making those range rates uncertain

Ampere's law: current passing through a conductor creates a magnetic field around the conductor; named after the man who discovered the phenomenon

Amplitude: the "height" of a trace on a display or record, usually referring to a measurement of voltage or power

Analog error signals: indicators derived from a continuous process that permits correction while the process is going on, as with a radar beam tracking a target

Analog signal processing: performing various operations on signals while they are still in their received form, that is, continuous rather than quantized or digitized

Analog-to-digital (AD) converter: electronic circuitry that samples an incoming signal and assigns a number (or a computer word) that describes each sample

Angle resolution: the capability of an antenna to separate two objects in angle

Angle tracking: to follow a target in angle, by whatever means

Angular error: the amount by which an antenna or other angle measuring system fails to indicate the exact angle of a target

Angular rate: the rapidity with which a target's position changes in angle

Antenna pattern: the "intensity" of the field at locations around the antenna at long distances from it; usually established by taking a series of measurements

Anti-jam gain or margin: the ratio by which a particular waveform or other technique is able to mitigate the effects of jamming

Aperture: literally an "opening"; its size determines the amount of electromagnetic energy intercepted; thus, any antenna is called an aperture

Arecibo: the location in Puerto Rico of a 1000-ft-aperture ionospheric and astronomic radar

Array factor: the component of an array antenna gain pattern due to the array of individual elements

Aspect angle: the angle made by some arbitrary characteristic of the target, say, its axis, with the axis of the radar antenna beam

Auroral ionosphere: the ionosphere near the Earth's magnetic poles when excited by magnetic storms

Azimuth: the angle from a fixed point, say, due north, in the plane of the Earth's surface (horizontal plane)

Barrage jamming: disrupting and interfering jamming that continuously covers all the radar frequencies being radiated

Beam steering: moving an antenna pattern about the sky, usually by electronic means

Beamwidth: the lateral dimension (in angle) of the principal lobe (main lobe or main beam) of an antenna pattern

Bistatic radar: a radar whose transmit and receive antennas are substantially separated

Blanking circuit: an electronic scheme by which particular range or angle locations in the radar coverage are wiped out

Blass array: an electronically steered array consisting of stacked beams that are turned on and off with a waveguide matrix; named after its inventor

Blind ranges: ranges where a filter that suppresses clutter at one range also suppresses signals at other ranges

Blind speeds: speeds at which a filter that suppresses clutter at one range rate also suppresses signals at other range rates

Boltzmann's constant: the number by which temperature (in kelvins) is related to energy (in joules or watt-seconds) per hertz of bandwidth; it is 1.38×10^{-23} joules/°K, named after the man who developed the phenomenon

Boresight: the forward direction on the axis of an antenna; loosely, the direction a reflector antenna is pointing or the direction perpendicular to the surface of any antenna

Butler array or matrix: an electronically steered array with the characteristics that there are as many beams formed as there are elements, they are orthogonal and the output is the Fourier transform of the input; named after its inventor

Carrier frequency: the rate of oscillation of the radio waves that carry a signal through space

Cassegrain feed: an antenna feed patterned after the optical telescope feed of the same name; the feed is located at the center of a parabolic reflector, reflecting onto the reflector via a hyperboloid at its focus

CFAR: constant false-alarm rate; keeping noise level constant by normalizing to a noise sample taken near the target

Chaff: small, light pieces of material that have high radar cross section

Circular polarization: a characteristic of radio waves whose electric vector rotates 360° during each radio frequency cycle

Clutter: unwanted and interfering radar returns from objects other than targets

Clutter coefficient: the ratio of the return received from clutter to what would have been received from a perfectly conducting isotropic radiator of the same physical area

Clutter fence: a screen around a radar site that prevents low-angle clutter sources from being illuminated

Coherence: the preservation of fixed phase relationships over time in the conduct of radar operations

Collision frequency: the rate at which the free electrons in an ionized medium collide with ions or atoms

Comb filters: an array of filters, arranged like the teeth of a comb, whose response frequencies are close together

Cone sphere: the shape created by a sphere capping a truncated cone; similar in shape to an ice cream cone

Conical scan: a tracking technique in which an antenna feed or the antenna itself makes a small, circular motion, sequentially comparing returns to obtain more accurate angle information about the target

Convolution: multiplying the overlapping portions of two functions continuously as one is moved across the other

Corner reflector (dihedral and trihedral): radar targets designed to have high retrodirective radar cross section. The dihedral has two faces, the trihedral three.

Cross range: a measurement orthogonal to the axis of an antenna, differentiated from arc length derived from an angle measurement

Data link: a communications channel for information, usually digitized and wide bandwidth

Dead zone: the region near a radar from which returns are not received because the receiver is either turned off or not connected to the antenna while the transmitter is radiating

Decibel: ten bels, a bel being the logarithm to the base ten of the ratio of output power to input power

Difference beam: the remainder or residue that results when the voltage from a signal in one beam is subtracted from that in an adjoining beam

Diffraction grating: a matrix of narrow slits that breaks light up into fringes caused by the constructive and destructive interference of light waves

Diffuse: to break up and distribute on reflection, as with an incident electromagnetic wave

Digital signal processing: performing various operations on signals after they have been converted to digital form, as differentiated from analog signal processing

Directional coupler: a switch that connects one electric circuit with another in such a way that energy moves easily in one direction but not in the other

Distributions: various probability density functions of mathematical statistics [exponential, gamma, Gaussian (normal), and Rayleigh] that find applications in radar theory

Doppler ambiguity: a condition in which range rate data folds over so that there is uncertainty about the true Doppler frequency

Doppler sidelobes: the residue of a filter's frequency response that appears in adjacent filters

Doppler spread: the band of frequencies within which Doppler returns might occur

Downrange: away from the radar along the axis of its antenna

EHF: extremely high frequency; frequency band

Electric vector: the direction and magnitude of the voltage measured in an electromagnetic field

Electron density: the number of free electrons per unit volume of an ionized gas or plasma

Electronic countermeasures (ECM): using electronic techniques to disrupt radar or communications

Electronic counter-countermeasures (ECCM): activities to counter ECM

Element factor: the pattern of individual elements of an array antenna

Envelope: the shape, amplitude, or modulation of a signal after the radio frequency carrier has been removed; the post-detection content of the signal

Envelope integration: building up the signal by summing two or more signal envelopes; post-detection integration

Erf(x): a form of the integral of the standard Gaussian distribution in which the standard deviation has been made narrower by the square root of two

ERP: effective radiated power of a radar or jammer (peak power × transmit antenna gain)

Ether: an imagined medium by which, it was thought until the nineteenth century, light waves were able to propagate in space

False-alarm probability: the likelihood that noise alone will cross a threshold and be erroneously accepted as a signal

False-alarm rate: the frequency with which noise alone crosses a threshold and is erroneously accepted as a signal

Fan beams: antenna patterns whose main lobes are in the shape of a fan

Faraday's law: a time-varying magnetic field will induce a voltage in a circuit immersed in that field; also known as the induction law; named after its discoverer

Faraday rotation: the turning of the electric vector of an electromagnetic wave as it passes through an ionized medium, an action that occurs due to Faraday's law

Far field: the region sufficiently far from an antenna that the phases of wavelets arriving from the antenna edges will be negligibly different from those arriving from the center; similar relationship for the radar cross section of a target

Fast Fourier transform (FFT): an algorithm for efficiently calculating the frequency content of a digitized time function

Fat beams: antenna patterns whose main lobes occupy a relatively large solid angle

Feedhorn: the expanding end of a waveguide that acts as a launcher of electromagnetic waves

FM chirp: a frequency-modulated signal that smoothly changes frequency upward or downward during its transmission; if it were a sound wave, it would be heard as a chirp

FM ramp: the ramp-like shape of the plot of a waveform whose frequency is changing either upward or downward during transmission

Fourier transforms: a mathematical operation that changes a function to reveal its characteristics in a different dimension; named after its inventor

Fraunhoffer diffraction: the far-field patterns that appear when light is propagated through a narrow slit or grating

Frequency ambiguity: uncertainty about the true Doppler shift of a target because of fold over in the Doppler processor

Frequency diversify: a characteristic of radars that can radiate at any one of a large number of frequencies; a design to reduce jamming vulnerability

Frequency domain: the dimension in which a function is evaluated for its spectral content

Frequency scanning: a technique by which an array antenna surveys a solid angle by changing its carrier frequency

Gain: the focusing power of an antenna as measured by the ratio of the angular area of a sphere to the angular area of the antenna beam; or the power of a processor to build up the signal-to-noise ratio by applying various techniques

Galactic noise: unwanted and interfering electromagnetic radiation that enters the radar from the cosmos

Gaussian: refers to the normal probability distribution; phenomena whose events are normally distributed are "Gaussian"

Geometric optics region: the region where the scattering of electromagnetic waves is from objects whose characteristic dimension is much larger than a wavelength

Geosynchronous: synchronized with the turning of the Earth; satellites in orbits whose period is 24 hr and whose inclination is zero degrees are geosynchronous

Gigahertz (GHz): billions (10^9) of cycles per second

Grating lobes: patterns in which energy appears at almost equal amplitudes in several locations, as with diffraction through multiple slits or interferometry

Grazing angle: the angle an antenna beam makes with the surface of the Earth

Gun-barrel analogy: the inference that the trajectories of radar targets can be compared with those of bullets emerging from a gun barrel

Half-wave dipole: a radiating or receiving element consisting of a straight conducting wire one-half as long as the wavelength of the associated radio waves

Hamming weighting: tailoring the amplitude of an antenna illumination function to reduce antenna sidelobes, or a waveform to reduce measurement sidelobes, so that it is the shape of a cosine on a pedestal; named after its designer

Haystack radar: a very large, high-precision reflector antenna radar atop Haystack Hill in Massachusetts

HF: high frequency, frequency band

Horizontal polarization: the condition of a radio wave whose electric vector is in the plane of the Earth's surface

Huggins beam steering: a method of moving a radar beam about the sky by adding and then subtracting the correct phases from the carrier frequency; named after its inventor

Hybrid: a device that employs a mixture of two or more electronic technologies, such as mixing tubes and solid state devices, analog and digital processing, and waveguide and electronic circuits

Hypothesis: test a method of decision making in mathematical statistics in which a threshold is set at a predetermined level, fixing the probability of incorrectly accepting or rejecting the hypothesis

Illumination function: the voltage or power pattern with which an antenna is excited

Noncoherent integration: the adding together of the envelopes or modulation of signals without regard for the phase of the carrier frequency; post-detection integration

Incremental sources: imaginary small radiators of electromagnetic energy used as a convenience for developing theory

Inertial guidance: a completely autonomous system of navigation that measures accelerations and deduces other quantities from those measurements and original position

Interferometer: a system that uses phase differences (constructive and destructive interference) to determine angular position to high accuracy

Interpulse period: the interval between radar pulses, pulse repetition interval (PRI), the reciprocal of the pulse repetition frequency (PRF)

Inverse synthetic aperture: the use of the deterministic rotation of an object in a radar beam to derive differential Doppler information and thereby resolve the object in angle; the "inverse" of having the radar beam pass across the target

Ionospheric radars: radars that irradiate the ionosphere so that scientific information can be derived from the noncoherent backscatter received

Ionospheric sounders: systems that probe the ionosphere by transmitting rapidly varying frequencies toward it; returns are evaluated to obtain estimates of electron density as a function of height

Isotropic radiator: an imaginary source of electromagnetic energy that radiates equal amplitudes and phases in all directions

Jitter: small, rapid, perhaps random fluctuations about an intended point or location

Kalman filter: a well known tracking algorithm (named after its creator) that weights measurements according to their quality to optimize results

Keplerian motion: movement dictated only by the forces of gravity, such as the movement of planets and satellites

Kilohertz (kHz): thousands (10^3) of cycles per second

Kilometers (km): thousands (10^3) of meters

Kilowatts (kW): thousands (10^3) of watts

Klystrons: high-power amplifiers of radio frequency energy, capable of coherent operation

Laser radar: a radar at optical, infrared or ultraviolet frequencies

Lidar: for "light detection and ranging"; a laser radar; sometimes, "ladar"

Linear array: an array antenna consisting of radiating elements arranged in a line

Line feed: a feed whose elements are arranged in a line; they may also be phased, as in a line feed that removes spherical aberration

LPI radar: low probability of intercept radar; a radar with waveform and antenna designed to minimize the power radiated in both spatial and spectral domains

Main beam: that part of an antenna pattern that contains the major portion of the energy

Matched filter: a filter in a radar receiver whose spectral response is matched to the spectral content of the transmitted waveform

Megahertz (MHz): millions (10^6) of cycles per second

Megawatt (MW): millions (10^6) of watts

Microseconds (μsec): one one-millionth (10^{-6}) of a second

Microwatt (μW): one one-millionth (10^{-6}) of a watt

Milliradian (mrad): an angle measure of one one-thousandth of a radian, that is, $0.0573°$

Millisecond (msec): one one-thousandth (10^{-3}) of a second

Milliwatt (mW): one one-thousandth (10^{-3}) of a watt

MMIC: monolithic microwave integrated circuit; refers to an all solid state array element integrated on a single piece of substrate

Modulation: the impression upon the radio frequency carrier of the signal fluctuations

Monopulse tracking: deriving out of a single pulse all the information necessary to obtain angle measurements

Moving target indication (MTI): the use of the Doppler content of the radar returns to achieve cancellation of clutter at zero range rate (or some other specific range rate)

MTI cancellers: the electronic circuits that accomplish the clutter cancellation for MTI

Newtonian trajectories: flight paths that have no forces acting on them except the forces of gravity

Noise: unwanted, sometimes random, electromagnetic energy that mixes with and interferes with the wanted signal energy

Normalize: to refer data to a convenient or common reference point and apply a standard interval from that point

North filter: a filter that gives an optimum response for signal against Gaussian noise, also called a *matched filter;* named after the person who first analyzed it

Nutate: to nod or wobble slightly

Nutation: a slow or small rotation superimposed on a more rapid or larger one

Orthogonal polarization: orientation of the electric field of electromagnetic radiation (including light) at right angles to a reference electric field

Oscillators: electric circuits that produce sinusoidal waves (waves that rise and fall smoothly and harmonically)

Over-the-horizon radar: radar that uses ionospheric reflection to detect targets beyond the horizon, operating in the frequency band that supports the phenomenon: the HF (high-frequency) band

Parabolic antenna: a device that radiates and focuses electromagnetic energy by use of the shape of the curve of a parabola

Paraboloid: a surface made up by revolving a parabola about its axis; the shape of a parabolic antenna

Parameter: an arbitrary constant that may take on or be assigned various values

Passive ECM: the use of such things as chaff and decoys to defeat radar or communications passively

Pedestal: the structure that supports a radar antenna

Pencil beam: electromagnetic energy focused to a narrow angle in two dimensions as with a searchlight beam

Phase array: an antenna that forms a beam by assigning phases to a number of separate radiating elements

Phase locked: phase held at a constant phase relationship by being tied electrically to a reference oscillator, usually a stable local oscillator

Phase shifter: electronic circuits that shift phase in discrete, predetermined steps

Plan view: the perspective from above

Plasma frequency: the rate of vibration of the electrons in an ionized medium

Polarization: the alignment of the electric field with respect to the propagation vector

Polarize: to align the electric vectors of electromagnetic radiation

Potted: physically fixed in position by being embedded in a resin or other seal

Power-aperture: the product of a radar's power and the physical aperture of its antenna

Power spectral density: the power present in the frequency constituents of a function (usually a plot of these quantities)

Propagation: the outward spreading of electromagnetic waves

Pseudorandom code: a series of random quantities (numbers or levels) whose randomness is unproven; or, random quantities that are replicated when generated for later autocorrelation

Pulse-burst waveform: a train of pulses

Pulse compression: a technique by which more bandwidth is inserted into a pulse than its duration would imply it could contain

Pulse Doppler: a radar or a waveform that uses a series of pulses that are processed for their range rate content

Quanta: small discrete packages of energy

Quantum mechanics: a theory of physics that treats the interactions of radiation and matter; its name derives from the observation that these interactions take place only in discrete packages

Radar: an instrument for radio detection and ranging

Radar altimeter: a radar that measures altitude

Radar cross section: a measure of the amount of electromagnetic energy a radar target intercepts and scatters back toward the radar

Radar signature: identifying features or patterns in a target's radar cross section

Radial velocity: the component of velocity vector on the radial toward or away from a point, for example, along the line of sight from a radar

Radian: form to express an angle in radians

Radian frequency: to express frequency in radians per second rather than in cycles or degrees per second

Random variable: in statistics, a function defined over a sample space; a nondeterministic variable

Range error: the inaccuracy in a range measurement

Range rate: the magnitude of the projection of the radar-to-target velocity vector (including both radar and target velocity vectors) on the radar-to-target range vector

Range rate ambiguities: foldover of range rates in the signal processor, requiring additional processing to determine the actual range rates of the targets

Range rate resolution: the ability to differentiate two targets in range rate

Range rate spectrum: the frequencies generated by a moving target

Range rate tracking: following a target in range rate

Range resolution: the ability to differentiate two targets in range

Range sidelobes: the residues of a pulse compression waveform that spill over into and contaminate adjacent range cells

Range tracking: following a target in range

Rayleigh region: a region where electromagnetic energy is scattering from targets that are smaller than a wavelength; named after Lord Rayleigh, who calculated the magnitude of that scattering

Rectification: processing of electric waves that swing both positive and negative into waves that swing only positive

Refraction: the bending of electromagnetic waves that takes place as the medium varies over the propagation path

Resolution bin or cell: the extent of the region (in angle, range, or range rate) filled by the return from a single-point target

Resonance region: the region where the wavelength of the scattered electromagnetic energy is of the same order as the characteristic dimension of the target

Root-mean-square (RMS): the square root of the average of the sum of the squares of a series of values

Scanning: moving a radar beam around the sky to cover a prescribed region

Scattering: the reflection of electromagnetic energy from a target

Scintillation: rapid variations in the level of scattering from a target

Semi-isotropic radiation: emission from a point source of equal levels of energy in all directions within a hemisphere

Servo drive: the power provided to equipment by a technique that uses system output to determine partially what the input will be

Servo loops: the electric circuits that sample system output and refer it back to the input

Servomechanism: an automatic device that uses feedback to control systems, usually by inserting at the input control signals derived from samples of the output

SHF: super high frequency; frequency band

Sidelobe: the unwanted, out-of-place residue of an antenna pattern or waveform

Sidelobe jamming: the act of sending interfering and disrupting signals into the radar antenna sidelobes

Side-looking radar: a radar that points at an angle substantially off the velocity vector of its carrier, hence, another name for a synthetic aperture radar

Signal-processing gain: the improvement in signal-to-noise ratio that results when various processing techniques are applied

Signal-to-noise ratio: the ratio of the RMS signal power to the RMS noise power at the output of a radar receiver

Sinusoid: sine or cosine plots

Skip distance: the range at which a radio wave propagated from the Earth toward the ionosphere returns to the Earth's surface

Solid angle: an area in angle, a number of square degrees or steradians

Spark gap: a mechanism by which electromagnetic energy is radiated by building up the field intensity across a gap until the intervening medium breaks down and a spark occurs

Specular returns: radar reflections of high amplitude and short duration, like flashes from a mirror

Spherical aberration: distortion in a wave front, resulting when it is reflected off a spherical surface rather than a parabola

Spherics: bursts of electromagnetic interference caused by disturbances in the atmosphere

Squint angle: the angle off the velocity vector of its carrier that a synthetic aperture radar may be pointed

STC: sensitivity time control: attenuating the radar return signal exponentially as a function of time to keep near-in returns from saturating the receiver

Steradian: the solid angle subtended by an area on the surface of a sphere equal to its radius squared

Sub-array factor: the component of an array antenna gain pattern due to a group of elements

Sub-cutter visibility: the capability of a radar processor to suppress clutter

Sum beam: the adding together of two slightly displaced antenna beams single beam that is the sum of the two, as with monopulse tracking

Surveillance: providing coverage of or keeping watch over

Synthetic aperture radar (SAR): a system that uses movement of an antenna beam across an area to synthesize a very large aperture and provide very good angle, and thus cross-range, resolution

Synthetic display: an uncluttered presentation obtained by distilling essential formation from noisy data and rejecting the latter

Tapering: varying the density of the elements in an array or the power by the array elements in order to obtain a tailored aperture illumination control the array far field pattern

Thermal noise: unwanted signals generated by the heat inherent in the of a radar system, a major factor in the first stage of radio frequency amplification

Threshold: a level established for decision-making as to whether or not a desired signal present

Time-delay networks: circuits in array radars that point the antenna various directions by progressively delaying the radiation of the signal array elements

Time domain: viewing a multidimensional function in its time dimension

Time sidelobes: the spreading residues in range of pulse compression forms, also called *range sidelobes*

Tracking: following selected targets over time, whether in range, angle, or range rate

Tracking gate: a region of special attention around a target being tracked with appropriate logic to keep the gate moving with the target

Track-while-scan (TWS): the radar operation in which targets are followed by the routine scanning function, differentiated from an operation where changes its routine to do tracking

Transponders: equipment that generates and radiates energy as a resulting a signal; it may reradiate an enhanced version of the signal received new information

Truncated cone: a conical shape that has been cut off at right angles to the axis of the cone at some arbitrary point

UHF: ultra high frequency; frequency band

Unambiguous range: the range associated with the time between radar pulses, that is, the maximum distance at which the round trip to a target can be completed before the next pulse is sent

Unambiguous range rate: a waveform design feature that provides that the range rates of the targets of interest will not fold over in the signal processor

Variance: a statistical quantity indicating the spread of a distribution about its mean

Vertical polarization: by convention, an electromagnetic wave whose electric vector is perpendicular to the Earth's surface

Vertical return: that part of an airborne radar signal that returns from directly beneath the aircraft where the angle of incidence is 90°

VHF: very high frequency; frequency band

Video integration: adding up signals after they have been through the envelope detector when only the envelope of the original signal remains

Waveforms: various shapes of radar pulses or groups of pulses designed to accomplish particular objectives

Weighting: changing the shapes of pulses or the envelopes of groups of pulses to tailor their sidelobes, usually by rounding the ends with a slowly varying function, such as a cosine

Woodward ambiguity function: the surface that results when responses to waveforms are mapped in both range rate and time; named after the person who led in analyzing ambiguity functions

Index

About the Co-author

PAUL J. HANNEN

Mr. Hannen has extensive experience in radar systems, electronic combat, survivability assessment, and modeling and simulation (M&S). Design and analysis experience in the following areas: mission analysis; radar systems; weapon delivery; survivability; electronic combat; design assessment; Kalman filters; and mission and defensive avionics for strategic, tactical, airlift, and reconnaissance aircraft. This experience has been applied to research and development, intelligence, acquisition, test and evaluation, and educational programs.

He is presently a Systems Engineer with Science Applications International Corp. (SAIC), Dayton, Ohio, and an Adjunct Professor with Wright State University. Previously, he worked as an electronics engineer with the USAF Avionics Laboratory, Wright-Patterson AFB, Ohio. Mr. Hannen received a BS (1979) and MS (1981) in Systems Engineering from Wright State.

Printed in the USA
CPSIA information can be obtained
at www.ICGtesting.com
JSHW051321221024
72173JS00006B/1279

9 781891 121340